T0094017

ALSO BY NEIL DEGRASSE TYSON

Letters from an Astrophysicist

Astrophysics for People in a Hurry

Space Chronicles: Facing the Ultimate Frontier

Origins: Fourteen Billion Years of Cosmic Evolution
with Donald Goldsmith

Accessory to War: The Unspoken Alliance Between Astrophysics and the Military
with Avis Lang

Cosmic Queries: StarTalk's Guide to Who We Are, How We Got Here, and Where We're Going
with James Trefil

Welcome to the Universe: A Pocket-Sized Tour
with Michael Strauss and J. Richard Gott III

STARRY MESSENGER

STARRY MESSENGER

Cosmic Perspectives on Civilization

NEIL DEGRASSE TYSON

A HOLT PAPERBACK
HENRY HOLT AND COMPANY
NEW YORK

Holt Paperbacks
Henry Holt and Company
Publishers since 1866
120 Broadway
New York, New York 10271
www.henryholt.com

A Holt Paperback® and H® are registered trademarks of Macmillan Publishing Group, LLC.

Distributed in Canada by Raincoast Book Distribution Limited

The Library of Congress has cataloged the hardcover edition as follows:

Names: Tyson, Neil deGrasse, author.
Title: Starry messenger : cosmic perspectives on civilization / Neil deGrasse Tyson.
Description: First edition. | New York : Henry Holt and Company, 2022. | Includes bibliographical references and index.
Identifiers: LCCN 2022026991 (print) | LCCN 2022026992 (ebook) | ISBN 9781250861504 (hardcover) | ISBN 9781250880949 | ISBN 9781250861498 (ebook)
Subjects: LCSH: Science and civilization. | Civilization—Philosophy. | Cosmology.
Classification: LCC CB478 .T96 2022 (print) | LCC CB478 (ebook) | DDC 901—dc23/eng/20220804
LC record available at https://lccn.loc.gov/2022026991
LC ebook record available at https://lccn.loc.gov/2022026992

ISBN: 9781250861511 (trade paperback)

Originally published in hardcover in 2022 by Henry Holt and Company

First Holt Paperbacks Edition 2024

Designed by Gabriel Guma

Printed in the United States of America

10 9 8 7 6 5 4 3 2 1

Dedicated to the memory of Cyril DeGrasse Tyson[1]
and all others who want to see the world
as it could be, rather than as it is.

You develop an instant global consciousness, a people orientation, an intense dissatisfaction with the state of the world, and a compulsion to do something about it.

From out there on the Moon, international politics look so petty. You want to grab a politician by the scruff of the neck and drag him a quarter of a million miles out and say, "Look at that, you son of a bitch."

—Edgar D. Mitchell, *Apollo* 14 astronaut

TABLE OF CONTENTS

PREFACE

Starry Messenger is a wake-up call to civilization. People no longer know who or what to trust. We sow hatred of others fueled by what we think is true, or what we want to be true, without regard to what is true. Cultural and political factions battle for the souls of communities and of nations. We've lost all sight of what distinguishes facts from opinions. We're quick with acts of aggression and slow with acts of kindness.

When Galileo Galilei published *Sidereus Nuncius* in 1610, he brought to Earth cosmic truths that had been waiting since antiquity to descend upon human thought. Galileo's freshly perfected telescope revealed a universe unlike anything people presumed to be true. Unlike anything people wanted to be true. Unlike anything people dared say was true. *Sidereus Nuncius* contained his observations of the Sun, Moon, and stars, as well as the planets and the Milky Way.

Two fast takeaways from his book: (1) human eyes alone are insufficient to reveal fundamental truths about the operations of nature, (2) Earth is not the center of all motion. It orbits the Sun as just one among the other known planets.

Sidereus Nuncius translates from the Latin to *Starry Messenger.*

These first-ever cosmic perspectives in our world were ego checks on our self-importance—messages from the stars forcing people to rethink our relationships to one another, to Earth, and to the cosmos. We otherwise risk believing the world revolves around us and our opinions. As an antidote, *Starry Messenger* offers ways to allocate our emotional and intellectual energies that reconcile with the biology, chemistry, and physics of the known universe. *Starry Messenger* recasts some of the most discussed and debated topics of our times—war, politics, religion, truth, beauty, gender, race, each an artificial battlefield on the landscape of life—and returns them to the reader in ways that foster accountability and wisdom in the service of civilization. I also intermittently explore how we might appear to space aliens who arrive on Earth with no preconceived notions of who or what we are—or how we should be. They serve as impartial observers of our mysterious ways, as they highlight inconsistencies, hypocrisies, and occasional idiocies in our lives.

Think of *Starry Messenger* as a trove of insights, informed by the universe and brought to you by the methods and tools of science.

STARRY MESSENGER

OVERTURE

SCIENCE & SOCIETY

When people disagree in our complex world of politics, religion, and culture, the causes are simple, even if the resolutions are not. We all wield different portfolios of knowledge. We possess different values, different priorities, and different understandings of all that unfolds around us. We see the world differently from one another, and by doing so, we construct tribes based on who looks like us, who prays to the same gods as we do, and who shares our moral code. Given the longtime Paleolithic isolation within our species, perhaps we should not be surprised by what evolution has wrought. Groupthink, even when it defies rational analysis, may have conferred survival advantages to our ancestors.[1]

If we instead back away from all that divides us, you might find common, unifying perspectives on the world. If so, watch where you step. That new vista is neither north nor south nor east nor west of where you stand. In fact, the place exists nowhere on the compass rose. One must ascend from Earth's surface to get there—to see Earth, and everybody on it, in a way that leaves you immune to provincial interpretations of the world. We speak of this transformation as the "overview effect," commonly experienced by astronauts who have orbited Earth. Add to this the discoveries of modern astrophysics as well as the math, science, and technology that birthed space exploration, and yes, a cosmic perspective is literally above it all.

Nearly every thought, every opinion, and every outlook I formulate on world affairs has been touched—informed and enlightened—by knowledge of our place on Earth and of our place in the universe. Far from being a cold, feelingless enterprise, there is, perhaps, nothing more human than the methods, tools, and discoveries of science. They shape modern civilization. What is civilization, if not what humans have built for themselves as a means to transcend primal urges and as a landscape on which to live, work, and play.

What then of our collective and persistent disagreements? All I can promise is that whatever opinions you currently hold, an infusion of science and rational thinking can render them deeper and more informed than ever before. This path can also expose any unfounded perspectives or unjustified emotions you may carry.

One can't realistically expect people to argue in the same way scientists do among themselves. That's because scientists are not in search of each other's opinions. We're in search of each other's data. Even when arguing opinions, you may be surprised how potent a rational perspective can be. When illuminated by it, you fast discover that Earth supports not many tribes, but only one—the human tribe. That's when many disagreements soften, while others simply evaporate, leaving you with nothing to argue about in the first place.

Science distinguishes itself from all other branches of human pursuit by its power to probe and understand the behavior of nature on a level that allows us to predict with accuracy, if not control, the outcomes of events in the natural world. Scientific discovery often carries the power to broaden and deepen perspectives on all things. Science especially enhances our health, wealth, and security, which are greater today for more people on Earth than at any other time in human history.

The scientific method, which underpins these achievements, is often conveyed with formal terms that reference induction, deduction, hypothesis, and experiment. But it can be summarized in one sentence, which is all about objectivity:

Do whatever it takes to avoid fooling yourself
into believing that something is true when it is false,
or that something is false when it is true.

This approach to knowing enjoys taproots in the eleventh century, as expressed by the Arabic scholar Ibn al-Haytham (AD 965–1040), also known as Alhazen. In particular, he cautioned the scientist against bias: "He should also suspect himself as he performs his critical examination of it, so that he may avoid falling into either prejudice or leniency."[2] Centuries later, during the European Renaissance, Leonardo da Vinci would be in full agreement: "The greatest deception men suffer is from their own opinion."[3] By the seventeenth century, shortly after the near-simultaneous inventions of both the microscope and the telescope, the scientific method would fully bloom, propelled by the work of astronomer Galileo and philosopher Sir Francis Bacon (Lord Verulam). In short, conduct experiments to test your hypothesis and allocate your confidence in proportion to the strength of your evidence.

Since then, we would further learn not to claim knowledge of a newly discovered truth until a majority of researchers obtain results consistent with one another. This code of conduct carries remarkable consequences. There's no law against publishing wrong or biased results. But the cost to you for doing so is high. If your research is checked by colleagues, and nobody can duplicate your findings, the integrity of your future research will be held suspect. If you commit outright fraud—if you knowingly fake data—and subsequent researchers on the subject uncover this, the revelation will end your career.

This internal, self-regulating system within science may be unique among professions, and it does not require the public or the press or politicians to make it work. Watching the machinery operate may nonetheless fascinate you. Just observe the flow of research papers that grace the pages of peer-reviewed scientific journals. This breeding ground of discovery is also, on occasion, a battlefield of scientific controversy. But if you handpick pre-consensus scientific research to serve cultural, economic, religious, or political objectives, you undermine the foundations of an informed democracy.

Not only that, conformity in science is anathema to progress. The persistent accusations that we take comfort in agreeing with one another come from those who have never attended scientific conferences. Think of such gatherings as "open season" on anybody's ideas being presented, no matter their seniority. That's good for the field. The successful ideas survive scrutiny. The bad ideas get discarded. Conformity is also laughable to scientists attempting to advance their careers. The best way to get famous in your own lifetime is to pose an idea that counters prevailing research and that earns a consistency of observations and experiment. Healthy disagreement is a natural state on the bleeding edge of discovery.

In 1660, a mere eighteen years after Galileo's death, the Royal Society of London was founded, and is still going strong as the world's oldest independent scientific academy. Newly advanced scientific ideas have been contested there ever since, inspired by its marvelously blunt motto, "Take nobody's word for it." In 1743, Benjamin Franklin founded the American Philosophical Society to promote "useful knowledge." They continue today in precisely that capacity, with members representing all fields of academic pursuit in both the sciences and humanities. And in 1863, a year when he clearly had more pressing matters at hand, Abraham Lincoln—the first Republican US president—signed into existence the National Academy of Sciences (NAS), based on an act of Congress. This august body would provide independent advice to the nation, founded in living memory, on matters relating to science and technology.

Into the twentieth century, a proliferation of agencies with scientific missions serves a similar purpose. In the US, these include the National Academy of Engineering (NAE); the National Academy of Medicine (NAM); the National Science Foundation (NSF); and the National Institutes of

Health (NIH). It also includes the National Aeronautics and Space Administration (NASA), which explores space and aeronautics; the National Institute of Standards and Technology (NIST), which explores the foundations of scientific measurement, on which all other measurements are based; the Department of Energy (DOE), which explores energy in all usable and useful forms; and the National Oceanic and Atmospheric Administration (NOAA), which explores Earth's weather and climate, and how they may impact commerce.

These centers of research, as well as other trusted sources of published science, can empower politicians in ways that lead to enlightened and informed governance. This won't happen until the people who vote, and the people they vote for, come to understand how and why science works. Scientific achievement among a nation's institutions of research constitutes the seedbed of that nation's future and is nourished by the breadth and depth of support the agencies may receive from the administrative bodies that govern them.

After thinking deeply about how a scientist views the world, about what Earth looks like from space, and about the magnitude of cosmic age and of infinite space, all terrestrial thoughts change. Your brain recalibrates life's priorities and reassesses the actions one might take in response. No outlook on culture, society, or civilization remains untouched. In that state of mind, the world looks different. You are transported.

You experience life through the lens of a cosmic perspective.

ONE

TRUTH & BEAUTY

Aesthetics in life and in the cosmos

Since antiquity, the subjects of truth and beauty have occupied the thoughts of our deepest thinkers—especially the minds of philosophers and theologians and the occasional poet such as John Keats, who observes within his 1819 poem "Ode on a Grecian Urn":[1]

Beauty is truth, truth beauty,—that is all

What might these subjects look like to visiting aliens who have crossed the Galaxy to visit us? They will have none of our biases. None of our preferences. None of our preconceived notions. They would offer a fresh look at what we value as humans. They might even notice that the very concept of truth on Earth is fraught with conflicting ideologies, in desperate need of scientific objectivity.

Endowed by methods and tools of inquiry refined over the

centuries, scientists may be the exclusive discoverers of what is objectively true in the universe. Objective truths apply to all people, places, and things, as well as all animals, vegetables, and minerals. Some of these truths apply across all of space and time. They are true even when you don't believe in them.

Objective truths don't come from any seated authority, nor from any single research paper. The press, in an attempt to break a story, may mislead the public's awareness of how science works by headlining a just-published scientific paper as the truth, perhaps also touting the academic pedigrees of the authors. When drawn from the frontier of thought, the truth still churns. Research can wander until experiments converge in one direction or another—or in no direction, a warning flag of no phenomenon at all. These crucial checks and balances commonly take years, which hardly ever counts as "breaking news."

Objective truths, established by repeated experiments that give consistent results, are not later found to be false. No need to revisit the question of whether Earth is round; whether the Sun is hot; whether humans and chimps share more than 98 percent identical DNA; or whether the air we breathe is 78 percent nitrogen. The era of "modern physics," born with the quantum revolution of the early twentieth century and the relativity revolution of around the same time, did not discard Newton's laws of motion and gravity. Instead, it described deeper realities of nature, made visible by ever-greater methods and tools of inquiry. Like a matryoshka nesting doll, modern physics enclosed classical physics within these larger truths. The only times science cannot assure objective truths is on the pre-consensus frontier of research. The only era in which science could not assure objective truths was before the seventeenth century, back when our senses—inadequate and biased—were the only tools at our disposal to inform us of the natural world. Objective truths exist independent of that five-sense perception of reality. With proper tools, they can be verified by anybody, at any time, and at any place.

Objective truths of science are not founded in belief systems. They are not established by the authority of leaders or the power of persuasion. Nor are they learned from repetition or gleaned from magical thinking. To deny objective truths is to be scientifically illiterate, not to be ideologically principled.

After all that, you'd think only one definition for truth should exist in this world, but no. At least two other kinds prevail that drive some of the most beautiful and the most violent expressions of human conduct. Personal truths have the power to command your mind, body, and soul, but are not evidence-based. Personal truths are what you're sure is true, even if you can't—especially if you can't—prove it. Some of these ideas derive from what you want to be true. Others take shape from charismatic leaders or sacred doctrines, either ancient or contemporary. For some, especially in monotheistic traditions, God and Truth are synonymous. The Christian Bible says so:[2]

> *Jesus saith unto him, I am the way, the truth, and the life: no man cometh unto the Father, but by me.*

Personal truths are what you may hold dear but have no real way of convincing others who disagree, except by heated argument, coercion, or force. These are the foundations of most people's opinions and are normally harmless when kept to yourself or argued over a beer. Is Jesus your savior? Did Muhammad serve as God's last prophet on Earth? Should the government support poor people? Are current immigration laws too tight or too loose? Is Beyoncé your Queen? In the *Star Trek* universe, which captain are you? Kirk or Picard—or Janeway?

Differences in opinion enrich the diversity of a nation, and ought to be cherished and respected in any free society, provided everyone remains free to disagree with one another and, most importantly, everyone remains open to rational arguments that could change your mind. Sadly, the conduct of many in social media has devolved to the opposite of this. Their recipe: find an opinion they disagree with and unleash waves of anger and outrage because your views do not agree with theirs. Social, political, or legislative attempts to require that everybody agree with your personal truths are ultimately dictatorships.

Among wine aficionados, there's the Latin expression, "In vino veritas," which translates to "In wine there is truth." Audacious for a beverage that contains 12 to 14 percent ethanol, a molecule that disrupts brain function and (irrelevantly) happens to be common in interstellar space. The epigram nonetheless implies that a group of people drinking wine will find themselves, unprompted, being calmly truthful with one another. Maybe that happens at some level with other alcoholic beverages. Even so, vanishingly few of us have ever seen a bar fight break out between two people drinking wine. Gin, maybe. Whisky, definitely. Chardonnay, no. Imagine the absurdity of such a line in a movie script: "I'm going to kick your ass, but only after I'm done sipping my Merlot!" The same incredulous claim can probably be said of marijuana. Smoking dens don't tend to be the places where fights break out. Supportive evidence, if cinematically anecdotal, that honest truth can breed understanding and reconciliation. Maybe that's because honesty is better than dishonesty, and truths are more beautiful than untruths.

Far beyond wine truths, and close cousins of personal truths, are political truths. These thoughts and ideas already resonate with your feelings but become unassailable truths from incessant repetition by forces of media that would have you believe them—a fundamental feature of propaganda. Such belief systems almost always insinuate or explicitly declare that who you are, or what you do, or how you do it, is superior to those you want to subjugate or conquer. It's no secret that people will give their lives, or take the lives of others, in support of what they believe. Often the less actual evidence that exists in support of an ideology, the more likely a person is willing to die for the cause. Aryan Germans of the 1930s weren't born thinking they were the master race to all other people in the world. They had to be indoctrinated. And they were. By an efficient, lubricated political machine. By 1939 and the start of World War II, millions were ready to die for it—and did.

The aesthetics of what is beautiful and desired in culture typically shifts from season to season, year to year, and from generation to generation, especially regarding fashion, art, architecture, and the human body. Based on the size of the cosmetics industry and the larger beauty industrial complex, visiting space aliens would surely think that we think we are ugly beyond repair, in persistent need of "improvements." We've designed household tools to straighten curly hair and to curl straight hair. We invented methods to replace missing hair and to remove unwanted hair. We use chemical dyes to darken light hair and to lighten dark hair. We don't tolerate acne or skin blemishes of any kind. We wear shoes that make us taller and perfumes that make us smell better. We use makeup to accentuate the good and suppress the bad elements of our appearance. In the end, there's not much real or chromosomal about our appearance. The beauty we've created is not even skin-deep. It washes off in the shower.

That which is objectively true or honestly authentic—especially on Earth or in the heavens—tends to possess a beauty of its own that transcends time, place, and culture. Sunsets remain mesmerizing, even though you get one every day. Beautiful as they are, we also know all about the thermonuclear energy sources in the Sun's core. We know about the tortuous journey of its photons as they climb out of the Sun. We know of their swift journey across space, until they refract through Earth's atmosphere, en route to my eye's retina. The brain then processes and "sees" the image of a sunset. These added facts—these scientific truths—have the power to deepen whatever meaning we may otherwise ascribe to nature's beauty.

Hardly any of us have ever grown tired of waterfalls or the full Moon ascending over a mountainous or urban horizon. We persistently fall speechless at the singular spectacle that is a total solar eclipse. Who can turn away from the crescent Moon and Venus, together, suspended in the twilight skies? Islam couldn't. That juxtaposition of a "star" with the crescent Moon remains a sacred symbol of the faith. Vincent van Gogh couldn't turn away either. On June 21, 1889,[3] he captured it from the pre-dawn skies in Saint-Rémy, France, creating what is perhaps his best-known painting, *The Starry*

Night. And we never seem to get enough landscape panoramas from planetary rovers or cosmic imagery delivered courtesy of the Hubble Space Telescope and other portals to the cosmos. The truths of nature are rampant with beauty and wonder, out to the largest of measures of space and time.

It's therefore no surprise that the God or gods we worship tend to occupy high places, if not the sky itself. Or we perceive high places as closer to God—from mountaintops to puffy clouds to the heavens. Noah's ark settled atop Mount Ararat, not on the edge of a lake or river. Moses didn't receive the Ten Commandments in a valley or on the plains. They came to him atop Mount Sinai. Mount Zion and the Mount of Olives are holy places in the Middle East, as is the Mount of Beatitudes, the likely location of Jesus's famous Sermon on the Mount.[4] Mount Olympus was a high place above the clouds, crowded with Greek gods. Not only that, altars tend to be built in high, not low, places, with Aztec human sacrifices, for example, typically held atop Mesoamerican pyramids.[5]

How often have we seen posters, or even fine art, depicting cherubs, angels, saints, or a bearded God himself floating on a cumulonimbus cloud—the greatest of them all. Cloud taxonomy fascinated the Scottish meteorologist Ralph Abercromby, and in 1896 he documented as many as he could around the world, creating a numerical sequence for them. You guessed it. Cumulonimbus clouds landed at number 9, unwittingly seeding the everlasting concept of being on "cloud nine" when in a blissful state.[6] Combine cloud nine with beams of sunlight reaching every corner of an image, and you can't help but think of divine beauty.

Animist religions, common to indigenous peoples around the world from Alaska to Australia, instead tend to assert that nature itself—the brook, the trees, the wind, the rain, and the mountains—is imbued with a kind of spirit energy. If ancient peoples had had access to the cosmic imagery of today, their deities might have enjoyed even more places of beauty to hang out in while looking over Earth. One nebula (PSR B1509–58), imaged by NASA's orbiting Nuclear Spectroscopic Telescope Array (NuSTAR) in x-ray light, resembles a huge glowing hand in space with a clearly visible wrist, palm, outstretched thumb, and fingers. Even though the nebula is the glowing remains of a dead, exploded star, that didn't stop people from dubbing it "The Hand of God."

Alongside their catalog IDs,[7] we typically name astrophysical nebulae for what they resemble, using all kinds of fun

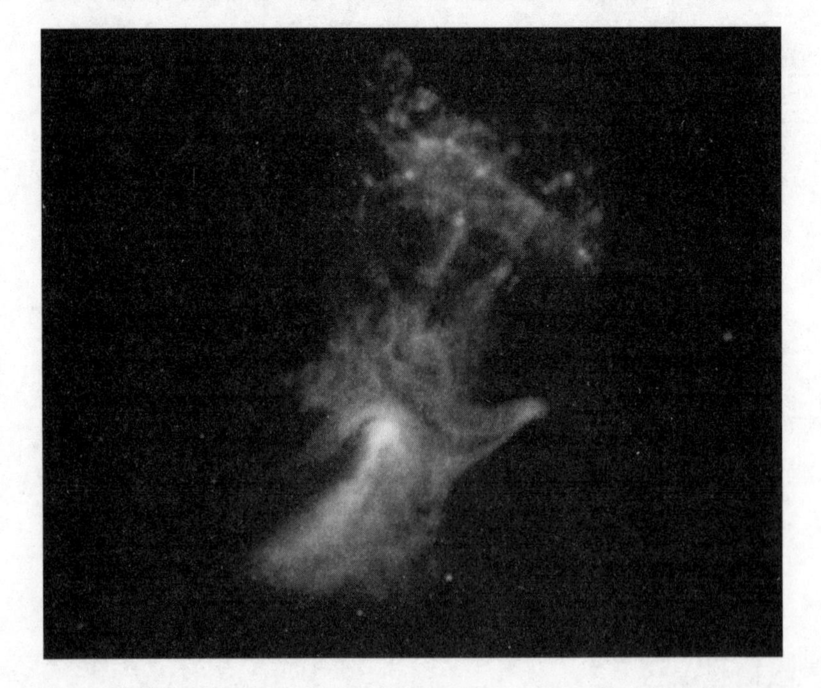

earthly references, including the Cat's Eye Nebula (NGC 6543), the Crab Nebula (NGC 1952), the Dumbbell Nebula (NGC 6853), the Eagle Nebula (NGC 6611), the Helix Nebula (NGC 7293), the Horsehead Nebula (IC 434), the Lagoon Nebula (NGC 6523), the Lemon Slice Nebula (IC 3568), the North American Nebula (NGC 7000), the Owl Nebula (NGC 3587), the Ring Nebula (NGC 6720), and the Tarantula Nebula (NGC 2070). Yes, they all actually look like or strongly evoke what we've called them. One more: the Pacman Nebula (NGC 281), named for the hungry 1980s video game character.

Splendor doesn't end there. In our own Solar System, we've got comets and planets and asteroids and moons, each revealing a stunning uniqueness of shape and form. For many of these objects, we've amassed intimate, objectively true knowledge of what they're made of, where they've come from, and where they're going. All while they rotate and move along their appointed paths through the vacuum of space, like pirouetting dancers in a cosmic ballet, choreographed by the forces of gravity.

In the White House of the 1990s, Bill Clinton kept on his Oval Office coffee table, between the two facing couches, a sample Moon rock brought back to Earth from a quarter-million miles away by Apollo astronauts. He told me that any time an argument was about to break out between geopolitical adversaries or recalcitrant members of Congress, he would point to the rock and remind people it came from the Moon.[8] This gesture often recalibrated the conversation, serving as a reminder that cosmic perspectives can force you to take pause and reflect on the meaning of life, and on the value of peace that sustains it.

A form of beauty unto itself.

But nature does not limit its beauty to things. Objectively true ideas can carry a beauty all their own. Allow me to choose some favorite examples:

One of the simplest equations in all of science is also the most profound: Einstein's equivalence of energy (E) and mass (m): $E = mc^2$. The small c stands for the speed of light—a constant that shows up in countless places as we unravel the cosmic codes that run the universe. Among a zillion other places that it shows up, this little equation underpins how all stars in the universe have generated energy since the beginning of time.

Equally simple, and no less profound, is Isaac Newton's second law of motion, which prescribes precisely how fast an object will accelerate (a) when you apply a force (F) to it: $F = ma$. The m stands for the mass of the object being pushed. This little equation, and Einstein's later extension of it from his Theory of Relativity, underpins all motion there ever was or will be for all objects in the universe.

Physics can be beautiful.

You've probably heard of pi—a number between 3 and 4 that harbors infinite decimal places, although often truncated to 3.14. Here's pi with enough digits to see all ten numerals 0 through 9:

$$3.14159265358979323846264338327950\ldots$$

You get pi simply by dividing the circumference of a circle by its diameter. That same ratio prevails no matter the size of the circle. The very existence of pi is a profound truth of Euclidean geometry, celebrated each year by all card-carrying geeks of the world on March 14—a date that can be written as 3.14.

Math can be beautiful.

Oxygen promotes combustion. Hydrogen is an explosive gas. Combine the two and get water (H_2O), a liquid that douses fires. Chlorine is a poisonous, caustic gas. Sodium is a metal, soft enough to cut with a butter knife and light enough to float on water. But don't try that at home because it reacts explosively in water. Combine the two and get sodium chloride (NaCl), more commonly known as table salt.

Chemistry can be beautiful.

Earth harbors at least 8.7 million species[9] of plants and animals, most of which are insects. This staggering diversity of life sprang forth from single-celled organisms four billion years ago. In this very moment a harmonic intersection of Earth's land, sea, and air supports every one of them. We are all in this together. One genetic family on spaceship Earth.

Biology can be beautiful.

What then of all that is true but ugly in the world? Earth is commonly thought to be a haven for life—nurtured by the maternal instincts of Mother Nature. That's true to an extent. Earth has been teeming with life ever since it could support life. Yet Earth is also a giant killing machine. More than 99 percent of all species that ever lived are now extinct[10] from forces such as regional and global climate change as well as environmental assaults such as volcanoes, hurricanes, tornadoes, earthquakes, tsunamis, disease, and infestations. The universe is also a killing machine, responsible for asteroid and comet impacts, the most famous of which struck Earth sixty-six million years ago, rendering all the famous oversized dinosaurs extinct, as well as 70 percent of all other land and marine species of life on Earth. No land animal larger than a duffel bag survived.

What's true but hard to admit is our morbid fascination with massive geologic catastrophes as well as destructive weather systems. They're all things of beauty—perhaps even an entire category unto itself: something to behold and admire, but only from a safe distance, although some people ignore the safe distance rule. How else do you breed "storm chasers" and death-wish meteorologists who report live from the docks while catastrophic storms batter the shoreline, drenching themselves and whoever was volunteered to hold the video camera that day.

A volcano is stunning at any angle. The red-hot fluid oozing from its caldera and down its slopes via tributaries and rivers is composed of liquefied rocks. At room temperature, these are things we sit on, build homes upon, and use as metaphor for all that is stable in the world. The volcano built itself with liquefied rocks, in that spot, on its own schedule, serving as a portal to Earth's literal underworld.

And is there anything more beautiful than a 300-mile-wide hurricane, viewed from on high or from space, slowly rotating like the gaseous pinwheel of storm clouds it is? How about a vigorous thunderstorm, with frequent, loud, and scary cloud-to-cloud and ground-to-cloud[11] lightning strikes?

And even though an asteroid took out Earth's big-toothed, badass dinosaurs, their absence pried open an ecological niche that allowed our tiny mammalian ancestors to evolve into something more ambitious than hors d'oeuvres for *T. rex*. That's undeniably a beautiful thing—at least for the branch on the tree of life that became primates, to which we belong.

~

Cosmic impacts can be destructive and deadly no matter where they occur. When sky-watchers Caroline and Eugene Shoemaker, along with David Levy, discovered comet Shoemaker-Levy 9 (one of many comets that bear their names), astro-geeks of the world all fought for a look through their telescope eyepieces. Why? After discovery, the comet's orbit was quickly determined to be on a collision course with the planet Jupiter. The world's astrophysicists mobilized our largest and most powerful telescopes, Hubble included. Previously scheduled observing slots were willingly forfeited. We even deputized *Galileo*, a Jupiter-destined space probe, not yet arrived, to join the observations. In a previous visit, Jupiter's strong tidal forces had ripped the comet apart, creating a parade of smaller chunks that maintained orbit. On July 16, 1994, we witnessed the first of nearly two dozen impacts—fragments A through W—on Jupiter. The biggest of these, fragment G, collided with the energy of six teratons (six million megatons) of TNT, equivalent to six hundred times the world's arsenal of nuclear weapons. These impacts left visible scars in Jupiter's atmosphere larger than Earth itself.

And it was beautiful.

A cosmic perspective cloaks the up-close damage and mayhem caused by these catastrophes. Their beauty subsumes all that is destructive. All that is lethal. Nothing died on Jupiter that day. Had those comet fragments collided with Earth, it would have been an extinction-level event.

Perhaps the line drawn between beautiful and ugly depends on whether it will harm us. Some objectively ugly things in nature might include a close-up of a tarantula's underbelly—lovable, perhaps, only to arachnologists. A tarantula can harm you with its bite, and maybe we know this intuitively. How about a Komodo dragon slowly stalking you? Or a swarm of bloodsucking ticks, or leeches? How about malaria? Or the bacterium that causes the bubonic plague? Or the virus that causes smallpox, or AIDS? How about all the spontaneous cell mutations that cause birth defects and cancers and other diseases that shorten our lives in the genetic lottery? They're all part of the same nature that contains countless objects and scenes we admire. But none of these parasites or diseases or creepy creatures show up in posters with Bible quotes. Smallpox, malaria, and the bubonic plague together have killed upwards of 1.5 billion people throughout time, worldwide. That toll far exceeds all deaths from all armed conflicts in the history of our species. Nature has killed more of us than we have of ourselves. These thoughts hardly ever (likely never) arise whenever we declare nature's beauty.

Maybe they should. If they did, we'd be more honest with ourselves about our place in the universe. Evidence shows that nature doesn't actually care about our health or longevity. We're equipped by natural instincts to sift between some of what might harm us and some of what may bring us comfort. Yet there is no hint from space that anyone or anything in the universe will arrive to save us from Earth, or from ourselves.

It is we alone who care about us.

Medical researchers develop vaccines to protect us from lethal viruses, and medicines to ward off bacteria and parasites. Architects and builders create homes and shelters to protect us from catastrophic weather. In the future, astrodynamicists will develop space systems that deflect the trajectories of killer asteroids headed our way. Contrary to implicit tenets of the green movement, not all that is natural is beautiful, and not all that is beautiful is natural.

Maybe that's why the world needs poets. Not to interpret what is plain and obvious, but to help us take pause and reflect on the beauty of people, places, and ideas—things we might otherwise take for granted. Simple beauty that emanates from simple truths. After reading Joyce Kilmer's most famous poem,[12] will you ever again walk past a tree without reflecting on its silent majesty?

I think that I shall never see
A poem lovely as a tree.

A tree whose hungry mouth is prest
Against the earth's sweet flowing breast;

A tree that looks at God all day,
And lifts her leafy arms to pray;

A tree that may in Summer wear
A nest of robins in her hair;

Upon whose bosom snow has lain;
Who intimately lives with rain.

Poems are made by fools like me,
But only God can make a tree.

Kilmer, a New Jersey native, was slain by a sniper's bullet on the Western Front in 1918, during World War I. One who died by the hands of another human and not by the hands of Mother Nature.

Where does this leave us? Perhaps nowhere. Perhaps everywhere. Personally, as a human, as a scientist, and as a resident of Earth, the most beautiful thing about the universe may be that it's knowable at all. No message written on tablets in the sky pre-required this to be so. It just is. For me, this summit of objective truth makes the universe itself the most beautiful thing in the universe.

TWO

EXPLORATION & DISCOVERY

The value of both when shaping civilization

Skeptics often think of space exploration as an expensive luxury, preferring instead to first solve our problems here on Earth. The list of societal challenges hasn't changed much over the decades and includes progressive goals to solve hunger and poverty, improve public education, reduce social and political unrest, and end war. These make potent headlines in any news cycle, but especially when contrasted with the tens of billions of dollars the US government spends annually in space. The topic is hotly debated in India,[1] a country that recently redoubled its efforts to explore space, all while eight hundred million of its citizens live in poverty. Half of those in poverty live in squalor[2]—more than the entire population of the US. Odd that these same skeptics hardly ever wonder whether we should do both: explore space and fix society's problems. The world's list of challenging problems long predates anybody ever spending a dime on space.

To gain insight, let's rewind thirty thousand years and eavesdrop on our ancestral cave dwellers. Those among them with the urge to explore decide to consult the elders, saying, "We want to see what's beyond the cave door." The elders are wise. They caucus among themselves, weighing what they think are the risks and rewards, and reply, "No. We must first solve the problems of the cave before anyone ventures beyond."

A laughably absurd exchange indeed, but to a space explorer, that's what people sound like when they require Earth's problems get solved before anybody goes anywhere else in the universe. One final note on our troglodytes: They breathed fresh air. They drank clean water. They ate organic plants and free-range animals—yet their staggeringly high infant mortality left them with an average life expectancy of barely thirty years. Modern science matters.

A cosmic perspective reminds you that Earth is a mote, isolated in a vast, rich universe. Is the cave any different from Earth itself? Yet, we knew more about the Moon before we visited for the first time than any fifteenth or sixteenth-century explorer knew about their destinations. We knew more about the surface of Mars and where to land our rovers than the twelfth and thirteenth-century Polynesian wayfinders knew about the Pacific islands that awaited them, far beyond their oceanic horizons. We spent centuries exploring and mapping Earth's surface, culminating with the discovery of Antarctica in 1820. Yet we've explored space for only a precious few decades.

If you travel beyond the cave door, you may just discover things that help solve your cave problems. To even suspect this is true requires enlightened foresight. You could find a diversity of plants that serve as medicine. You could discover an assortment of materials—wood, stone, bone—useful to fashion into tools. You could reveal additional sources of water, food, and shelter. More importantly, perspective lurks beyond the cave door. These places are not only destinations but new ways of seeing things. You don't need a scientist to tell you that. The noted writer T. S. Eliot once mused,[3]

> *We shall not cease from exploration*
> *And the end of all our exploring*
> *Will be to arrive where we started*
> *And know the place for the first time.*

That's as close an analogue to the cosmic perspective as has ever been penned by a poet.

Part of the problem is that we're all wired with linear minds, leaving us prone to think small. It's not our fault. We think in additives and multiples, with no evolutionary pressure to think in exponentials. An exponential is a number raised to the power of another number. When you do that, quantities and rates described by them rise (or shrink) faster than our normal capacity to comprehend. Consider this simple example: You can choose to receive $5 million now or instead receive a penny a day doubled for a month. Most people would take the $5 million and run, avoiding the pennies altogether. Let's first think it through. That's a penny today. Then two pennies tomorrow. Four pennies the next day. Eight pennies the day after that, and so on. How rich are you at the end? If you do the math, on the thirty-first day you will be handed $10,737,418.24. And the sum of pennies from your previous thirty days brings your total to $21,474,836.47. That's the power of an exponential.

In one more example, you learn that a species of unwanted

algae is spreading on the surface of your favorite pond. The growth is persistent and doubles in area every day. After a month, the lake is half-covered in algae. At this rate, how much longer until algae covers the entire lake? Our primitive, linear brain calculates "one month." But the actual answer is "one day." Doesn't even matter how long it took for the pond to become half-covered. If the rate of coverage doubles every day, you can be sure that when it's half-covered, you've got just one day left.

The devastating collapse of our economic system in 2008 was precipitated by predatory low-interest, floating-rate loans that granted mortgages to unqualified people. Who knows whether this economic episode could have been softened, or avoided altogether, if the people whose loans were approved were also fluent in exponential calculations. They would have realized in an instant that any uptick in compounded interest rates would leave them bankrupt, empowering them to decline the loan in the first place.

Consider how often we do simple linear calculations in our head: We've been driving for an hour and we're halfway there, so there's one more hour before we're home. That's straightforward linear thinking. But, in the spirit of "Are we there yet?" here's a sentence that has never been uttered in the history of transportation:

> *We've been driving for a thousandth of a second and we're one three-millionth of the way there, so just 2,999.999 more seconds before we're home.*

Yet mathematical factors of millions and billions and trillions are cosmically commonplace. The sphere of the Earth, which is so large that some people (still) think it's flat, is dwarfed by the Sun. If the Sun were hollow, you could pour a million Earths into its volume and still have room left over. Let's not stop there. In five billion years, when the Sun dies, it will pass through a phase called Red Giant, in which it swells enormously, engulfing the orbits of Mercury, Venus, and likely Earth as well. At that size, the Sun has ballooned to ten million times larger than it is now. The Solar System, out to the Kuiper Belt of comets beyond Neptune, is a million times larger still. The soul of the cosmic perspective—its spirit energy—derives from embracing these astronomical scales of measure. An inability to do so can thwart attempts to fathom the depths of time through which we live, and space through which we move.

You also need measurement moxie to embrace modern biology and geology. We think of Darwinian evolution as imperceptibly slow. That's because we live, at most, 100 years, and our brain wiring resists the fact that speciation can take thousands and even millions of times longer than our lifetime to unfold. That's how you go from our ancestral mammalian rodents running underfoot of *T. rex* to human beings in 66 million years—a stretch of time just 1.5 percent of the 3.8 billion years that Earth has hosted life. Still feels like long ago? Know what also takes a long time? The geologic carving of landforms such as Arizona's Grand Canyon, and continental drift, where Earth's largest landmasses move across the surface at about the same rate your fingernails grow. How about my favorite? If the gridiron of a football field were a timeline of the universe, with the Big Bang at one end and this moment at the other, then all of human recorded history would span the thickness of a blade of grass in the end zone.

Terrestrial exploration and discovery have long been associated with land grabs by military or colonizing powers, epitomized by Julius Caesar's infamous Latin quip (c. 45 BC):

Veni, vidi, vici (I came, I saw, I conquered).

The activity also involved flag planting on uncharted, never-before-visited places, such as the South Pole or the summit of Mount Everest. Flag planting also occurred where local residents were already there to greet you, which brings to mind what Christopher Columbus wrote to King Ferdinand and Queen Isabella in 1493 after his first voyage to the Caribbean:[4]

I discovered many islands inhabited by numerous people. I took possession of all of them for our most fortunate King by making public proclamation and unfurling his [flag].

Even the *Apollo 11* mission to the Moon planted a flag—the American flag. Although the plaque that accompanied it was unlike any other in the history of hegemony:

HERE MEN FROM THE PLANET EARTH
FIRST SET FOOT UPON THE MOON
JULY 1969, A.D.
WE CAME IN PEACE FOR ALL MANKIND

With Earth charted and the Moon visited, our collective concept of exploration and discovery must now extend further in the Solar System and beyond. The exercise also includes the discovery of ideas and inventions and new ways of doing things.[5] With systems in place to disseminate thought, such as scientific conferences, peer-reviewed journals, and patent filings, every next generation can use discoveries of the previous generation as fresh starting points. No reinventing the wheel. No wasted efforts. This blunt and obvious fact carries profound consequences. It means knowledge grows exponentially, not linearly, rendering our brains hopeless in our attempts to predict the future based on the past. It also leaves you thinking that all the amazing discoveries and inventions—the ones in your lifetime—mean you live in special times. Yet that's a fundamental feature of exponential growth: everyone thinks they live in special times, no matter where they are on the curve. How often have we all heard the phrase, "the miracles of modern medicine"? Now look back fifty years at the doctor's bag with scary tools and questionable cures and you take smug delight in being alive today instead of at any other time. People back then also praised their own state of advances relative to fifty years earlier. At no time on the exponential growth curve did anybody say, "Gee, we sure live in backward times," no matter how backward it may look to subsequent generations.

Borrowing the math from our pennies and algae, what's the "doubling time" of exploration and discovery? In 1995, while a postdoc at Princeton University, I decided it might be fun to measure a wall of research journals on the shelves of the Peyton Hall astrophysics library. A single publication, *The Astrophysical Journal*, is preeminent in my field and occupied most of the shelves. Perfect setup for my doubling experiment. The inaugural journal dates from 1895. All I did was find the middle of the wall and record the year of the journals at that location. It was 1980. That meant there was as much published astrophysics research in the fifteen years from 1980 to 1995 as had been previously published since 1895. That's a fifteen-year doubling time, but would it persist back to the beginning? I then found the midway point between 1895 and 1980. It was 1965. The next midpoint was 1950, followed by 1935, and then 1920. I may be off by a year or two because over time the *Journal* increased its page size. I was measuring shelf space when, to be precise, I should have been summing printed page areas, but the lesson of this exercise is nonetheless clear.

You might think the emergent publish-or-perish culture of academia has increased pressure to generate frivolous papers, artificially boosting the researcher's productivity. No. It's driven by the sheer increase in the number of researchers and the productivity that comes from large collaborations.[6] I would come to learn that a fifteen-year doubling time is consistent with the pace of other active fields of scientific research.

How about inventions? The US Patent and Trademark Office registered 3.5 million patents from 2010 to 2020, more than was registered in the nearly forty-year stretch from 1963 to 2000.[7] So they're on a roll too.

All this had me wondering: What might be the doubling time of modern society? And how would you measure it? I don't know and I don't know, but I'm happy to try. Let's look at thirty-year runs of the industrialized world since 1870, with emphasis on the United States, and compare life at the beginning to life at the end of each interval. How have exploration and discovery, the drivers of science, shaped our lives?

From 1870 to 1900, there are great advances in transportation. Steamships cross the oceans in record times. In 1869 the last "golden spike" is hammered, completing the 2,000-mile transcontinental railroad across the US. This enables decades of mobility and expansion for the population. In 1893 the legendary Orient Express, among many rail routes on the continent, begins its 1,400-mile circuit between Paris and Istanbul. Rail travel rendered transportation by stagecoach obsolete along many routes. Also, in the 1880s, German engineer Karl Benz improves on the internal combustion engine and births the first practical automobile. English inventor John Kemp Starley perfects the velocipede[8]—credit him for the now-familiar "safety bicycle," which uses two wheels of equal size and a chain connecting the rear wheel to the pedals. And over that time, powered balloons that enable transportation through the air are all the rage.

Daily life in 1900 would be unrecognizable to anyone transported from the year 1870.

Time to look at published predictions in 1900 for the year 2000. That's what people do when a new century begins. With the clever subtitle "The History of the Future," the website paleofuture.com specializes in just that. Rampant among predictions, such as what appeared in the publications *Punch*, the *Atlantic Monthly*, and *Collier's*, are simple linear extrapolations of what was happening in 1900. They see the promise of electric lighting but imagine it only for special occasions. They love airship travel and imagine that everyone in the future moves around via their own private balloon, including Santa—because who needs magic reindeer when you have blimps. Again, humans are linear thinkers, so you can't blame any of them for these quaint imaginings of their future.

The *Brooklyn Daily Eagle*'s last Sunday newspaper of the nineteenth century included a sixteen-page supplement of articles and illustrations titled, "Things Will Be So Different a Hundred Years Hence." The contributors—business and military leaders, pastors, politicians, and other experts in their fields—opined on what housework, poverty, religion, sanitation, and war would be like in the year 2000. They enthused about the potential of electricity and the automobile. There is even a map of the world-to-be, showing an American Federation comprising most of the Western Hemisphere from the lands above the Arctic Circle down to the archipelago of Tierra del Fuego, plus sub-Saharan Africa, the southern half of Australia, and New Zealand.

Most of the writers portray a future rich with fanciful extensions of the day's technologies, although one futurist could not see the future at all. George H. Daniels, who worked for the New York Central and Hudson River Railroad, peered into his own crystal ball and predicted,

> *It is scarcely possible that the twentieth century will witness improvements in transportation that will be as great as were those of the nineteenth century.*

Written just three years before the invention of flight, that's gotta be the most boneheaded prediction ever made. Rather than simply under-predict the future, like everybody else, he actively denies a future of innovations—in his own field. Elsewhere in his article, Daniels envisions affordable global tourism and the diffusion of white bread to China and Japan. Yet he simply can't imagine what might replace steam as the power source for ground transportation, let alone a heavier-than-air vehicle flying through the air. Even though he stood on the doorstep of the twentieth century, this manager of the world's biggest commuter rail system could not see beyond the automobile, the locomotive, and the steamship. Yet another victim of linear thinking unwittingly embedded in exponential growth.

Between 1900 and 1930, the existence of atoms is confirmed. Powered "aero"planes are invented, and the range of flight extends from the 120-foot (36-meter) distance flown in 1903 by the Wright brothers in their original Wright Flyer, to a 5,218-mile closed-circuit trip[9] in 1930, logged by the Italian aviators Major Umberto Maddalena and Lieutenant Fausto Cecconi. Back on the ground, we learn to exploit radio waves as a fundamental source of information and entertainment. Urban transportation shifts almost entirely from a horse-driven economy, the backbone of civilization for thousands of years, to an automobile economy, in which you can't give away a horse. This period also sees a world war, in which planes are used in combat for the first time. Orville Wright, writing from Dayton, Ohio, laments this fact in a letter dated December 19, 1918, to Alan R. Hawley, President of the Aero Club of America, New York City:[10]

> *Many thanks for your telegram remembering the fifteenth anniversary of our first flight at Kitty Hawk. Although Wilbur, as well as myself, would have preferred*

> *to see the aeroplane developed more along peaceful lines, yet I believe that its use in this great war will give encouragement for its use in other ways.*

Meanwhile, cities are electrified. To read at night, no longer do you burn wax, whale oil, or any other source of flame. And over this time cinema, silent and in black and white, becomes a leading source of entertainment.

Daily life in 1930 would be unrecognizable to anyone transported from the year 1900.

From 1930 to 1960, we go from airplanes flying at speeds of a few hundred miles per hour, to breaking the sound barrier in 1947, to the dawn of the space age, inspired in part by ballistic rockets developed as wartime weapons. In 1957 the Soviet Union launches Sputnik, Earth's first artificial satellite, which travels at 17,500 mph in low-Earth orbit. In 1958, the world's first commercial jet airplane—the Boeing 707, flown by Pan American Airways—enjoys a wingspan wider than the distance flown by the Wright brothers' first flight in 1903. This period also sees another world war and the invention of the laser. Nuclear weapons, from their invention in 1945 to 1960 (a mere fifteen years), increase in destructive power by nearly a factor of 4,000, accompanied by rocket and suborbital missile technologies to deliver their destructive power anywhere on Earth's surface within forty-five minutes. We see the rise of television as a potent source of instant information and entertainment, as well as the further rise of cinema, now in color and with sound.

Daily life in 1960 would be unrecognizable to anyone transported from the year 1930.

From 1960 to 1990, a Cold War nuclear arms race between the United States and the Soviet Union threatens the survival of civilization. Though begun in the 1950s, the US stockpile of nuclear warheads peaks in the 1960s, with the Soviets' stockpile peaking in the 1980s.[11] The Berlin Wall, erected in 1961, becomes the greatest symbol of Winston Churchill's "Iron Curtain," separating Eastern from Western Europe. Yet it's dismantled by 1989, as peace breaks out in Europe. The commercialization of the transistor allows consumer electronics to miniaturize, transforming audiovisual equipment from heavy, floor-mounted living room furniture to what you carry in your pocket. The laser goes from a specialized piece of laboratory equipment costing tens of thousands of dollars to a $5.99 laser pointer impulse buy at the checkout line of Walmart. Women enter the workforce in huge numbers, especially in professional fields traditionally held by men. Sunday newspapers rebrand their "Women's Section" as the "Home Section." The modern gay rights movement catapults to mainstream attention via the AIDS epidemic that sweeps the world. Not until 1987 is homosexuality removed entirely from the formal list of mental "disorders" compiled by the American Psychiatric Association.[12] Computers go from being expensive, room-size, special-purpose machines used exclusively by the military and by scientists to desktop necessities. The personal computer, introduced in the 1980s by IBM and Apple Computer, permanently transforms the daily habits of how people work and play. Hospitals in the 1980s see widespread use of MRIs, a potent tool in the medical professional's arsenal when diagnosing the condition of the human body without first cutting you open.

In *Back to the Future Part II* (1989), the filmmakers imagine life in the distant future of 2015. Flying cars are there. Everybody wants flying cars in the future. But in one scene, Marty McFly, the main character, angers his boss during a video call at home, and gets fired from his job. This bad news is express-delivered in that moment via fax. Not across just one fax machine. His futuristic residence has three, because if everyone owns one fax machine in 1989, then in the future, twenty-six years later, everyone would surely own three of them. In all fairness to Hollywood, it wasn't just the movies. In 1993 AT&T launched an ad campaign about the future, taglined "You Will." It got most stuff right, but included a TV spot with a person in a reclined chair on a seashore, scribbling on a tablet, about to do something I've never wanted to do, have never needed to do, have never done, nor ever will do. The voice-over boasts: "Have you ever sent someone a fax . . . from the beach? You will. And the company that'll bring it to you: AT&T."

Just one more thing. Between 1960 and 1990, we build the most powerful rocket ever to launch, and with it journey nine times to the Moon. While there we orbit, land, walk, skip, golf, and drive three electric buggies across its barren dusty terrain. We also develop a reusable space shuttle and allocate funds to build an international orbiting space station the size of a football field. Sorry, that's three more things.

Daily life in 1990 would be unrecognizable to anyone transported from the year 1960.

From 1990 to 2020, we map the human genome. Computers become portable—small enough to carry in a backpack. The World Wide Web, invented in 1989 by British computer scientist Tim Berners-Lee at the European Organization for Nuclear Research (CERN) in Switzerland, becomes ubiquitous in the 1990s. By 2000, searchable websites and e-commerce are commonplace, and everyone with a computer and access to the internet obtains an email address. Early in this period, mobile phones had become standard for anyone leaving the house, but beginning in 2007 are rapidly replaced by the pocket-size smartphone, granting full access to music, media, and the internet. Smartphones further host countless utilities that enhance everyday life, including a camera that shoots high-quality photos and video. Oh, and it also makes phone calls. The smartphone may be the single greatest invention in the history of inventions. In 2020 there are three billion of them in a world of eight billion people. Before 2007 there were zero. Show a smartphone to anyone in 1990, and they may just resurrect witch-burning laws to eradicate your magic.

In 1996, Global Position Satellites (GPS), a navigation tool created exclusively by the US military for national security, is formally opened to commercial interests. That's when President Bill Clinton issues a policy directive declaring GPS to be a dual-use system. Navigation tools rapidly commercialize to serve everything from tracking packages to hailing a car service to selecting a mutually interested sex partner within four blocks of your location. In the 2000s social media platforms change the communication landscape of family, friends, and especially politics. You can now drive across the US in an electric car, charging at any of forty thousand power stations along the way, all while we glimpse the dawn of electric self-driving cars. A few more things. Once ubiquitous among storefronts, video stores are gone. CDs and DVDs come and go. Computers are now smart enough to beat all humans in the board games chess and Go, the TV game show *Jeopardy!*, and damn near everything else that requires brain power. And this sentence, with clear and present meaning in 2020:

Google it to see if there's a smartphone selfie video posted to YouTube in 4K that went viral.

. . . is laden with mysterious nouns and verbs with no meaning to anyone in 1990.

You know you're living in the future when you can board a hundred-ton pressurized aluminum tube with wings, fly smoothly in a cushioned chair at 500 miles per hour, 31,000 feet above Earth's surface, and while crossing the continent, get served a pasta dinner and a mixed drink by someone whose job, in part, is to make you comfortable. And for most of the trip you surf the internet, watch any one of a hundred movies, only to land safely and smoothly a few hours later and complain that the marinara sauce was not to your liking.

With that run of societal advances, small wonder that in places such as the US, the wisdom of elders carries only marginal sway, which accounts for much of the tension that roils multigenerational Thanksgiving dinners. Their advice is predictably out of touch for what your college major should be, what jobs you should seek, what cars you should buy, what medicines you should take, what jokes you should tell, and what foods you should eat. Unless you happen to live in one of the world's coveted "Blue Zones" where people routinely live to 100 years old. If so, then just do everything those elders tell you to do, especially since they (probably) don't live in caves. In any case, your elders may carry more wisdom than you for navigating human sentiments such as love, kindness, integrity, and honor, which remain among the few constants of the world.

Today swarms with rampant predictions for what we can expect mid-century—the year 2050. If things continue as they have for the past 150 years, you can guarantee every one of them will fail. Maybe that's good because lately, most predictions are bleak. Many foretell a climate-change apocalypse. Some fear a virus catastrophe with lethality far greater than the six million deaths wrought by the COVID-19 pandemic of 2020. Others fear that artificial intelligence will escape its virtual box and ultimately become our overlord. On TV and in the movies, the zombie apocalypse feels real. A fan once asked twentieth-century sci-fi novelist Ray Bradbury why he imagined bleak futures. Is civilization doomed? He replied, "No. I write about these futures so that you know to avoid them."[13]

So when someone hints or declares they have any clue what the state of the world will be in the year 2050—thirty years from 2020—I reflect on our parade of thirty-year windows on human affairs. There's nothing linear about any of it. The river of discoveries in the natural world grows exponentially, fed by emergent tributaries of insight and knowledge, guaranteeing to embarrass any futurist. But that won't stop me from trying—with the caveat that when I first saw the *Star Trek* TV series in the 1960s, I fully embraced a future of warp drives, photon torpedoes, phasers, transporters, and aliens, but I remember thinking, "No, a door can't possibly know to open for you when you approach it." So, time to make a future-fool of myself:

By the year 2050 . . .

- Neuroscience and our understanding of the human mind will become so advanced that mental illness will be cured, leaving psychologists and psychiatrists without jobs.

- In a shift that echoes the rapid conversion from horses to automobiles in the early twentieth century, self-driving electric vehicles will fully replace all cars and trucks on the road. If you want to be nostalgic with your fancy combustion-engine sports car, you can drive on specially designed tracks, akin to horse-riding stables of today.

- The human space program will fully transition to a space industry, supported not by tax dollars but by tourism and anything else people dream of doing in space.

- We develop a perfect antiviral serum and cure cancer.

- Medicines will tailor to your own DNA, leaving no adverse side effects.

- We will resist the urge to merge the circuitry of computers with the circuitry of our brains.

- We will learn how to regrow lost limbs and failing organs, bringing us up to the level of other regenerating animals on Earth, like salamanders, starfish, and lobsters.

- Instead of becoming our overlord and enslaving us all, artificial intelligence will be just another helpful feature of the tech infrastructures that serve our daily lives.

What's driving it all? When we think of civilization, we commonly reflect on how engineering and technology have shaped our lives. Dig a little deeper and find ongoing scientific discovery that enables and empowers this progress. Nineteenth-century advances in thermodynamics gave engineers the necessary understanding of energy and heat to design and perfect their engines. Around the same time, discoveries in electromagnetism informed all thoughts about how to create and distribute electrical power. Einstein's Theory of Relativity, published in 1905 and 1915, would ultimately provide the precision that GPS satellite timing requires, among countless other extraordinary revelations about our universe, from how the Sun makes energy to the Big Bang itself. Quantum mechanics in the 1920s became the landscape on which all modern electronics would be based, especially the creation, storage, and retrieval of digital information. Material science, an ongoing enterprise that explores new alloys, composites, and textures of surfaces, has transformed everything we see, touch, wear, and use in our industrialized world. Each of these areas represents entire branches of the physical sciences with discoveries reported by scientists in research papers that once appeared in journals on shelves, but now display on the internet you didn't have in 1990.

Yes, we live in special times, only because it's all special, which brings to mind this oft-cited though shortsighted verse from Ecclesiastes, written thousands of years ago:[14]

The thing that hath been, it is that which shall be; and that which is done is that which shall be done: and there is no new thing under the sun.

You can only conceive this sentence in a pre-scientific era—before exponentials had even been imagined, and before anybody emerged from the dark of the cave to explore. Today, it's all new under the Sun—and Moon and stars. The only thing about it that doesn't change is the exponential rate of change itself.

THREE

EARTH & MOON

Cosmic perspectives

Twenty-seven astronauts, via the Apollo program's mighty *Saturn V* rockets, left Earth for the Moon—a quarter-million miles away. Except for those few who serviced the Hubble Space Telescope, and some SpaceX tourist missions, the remaining five hundred astronauts who have orbited Earth did not ascend much higher than about 250 miles above sea level. A distance rivaling that from Paris to London. Or from Islamabad to Kabul. Or from Kyoto to Tokyo. Or from Cairo to Jerusalem. Or from Daegu to Pyongyang. If you're a geographically challenged American, then it's slightly farther than from New York City to Washington, DC. Next time you hold a schoolroom globe of Earth in your hand, have a look at the distances between any of these pairs of cities. They sit about a centimeter from each other. Which means, what we've been calling "space travel" all these years has been astronauts orbiting one centimeter above a schoolroom globe, boldly going where hundreds have gone before, at a distance you can drive in less than four hours if you headed straight up and obeyed terrestrial speed limits.

My low-earth-orbit smackdown notwithstanding, ascending even these modest distances can offer enlightened perspectives. When not viewed through binoculars or high-resolution cameras, very little of human civilization is recognizable from orbit. Yes, the lights of cities at night can be striking from above, although not much more striking than from an airplane. No, you won't see the Great Wall of China, nor the US interstate system, itself wider than the Great Wall, and you will just barely make out the Hoover Dam and the Great Pyramid in Egypt. The resolution of the human eye-brain combination is about one arc minute. The highly touted Apple "retina" display was their successful attempt to create pixels smaller than your eye can resolve—smaller than one arc minute in your field of view at the typical distance you would use the display. With that resolution, the eye-brain system sees no pixels, bringing unprecedented clarity and sharpness to images. With 20/20 vision, that's equivalent to the size of a walnut across the length of a football field. The Hubble Telescope, which has better than 20/20 vision, and which has military versions of itself looking downward instead of upward, can resolve a walnut at a distance of one hundred miles.

With only one or two human-made structures visible from Earth orbit, everything else that divides us—national borders, politics, languages, skin color, who you worship—is invisible to you. The color-coded countries painted on our maps offer stark reminders of who is not us, thus helping to identify our allies as well as our enemies.

Astronaut Mike Massimino, an engineer whose Space Shuttle missions included servicing the Hubble Space Telescope, writes poignantly in his memoir about viewing Earth from space.[1]

> *The thought that went through my mind when I spacewalked and looked down on Earth was that this must be the view from Heaven. Then it was replaced by another thought: "No, this is what Heaven must look like."*

Mike gets misty-eyed every time he shares that experience. He went a little higher than the International Space Station. The Hubble orbits at 340 miles up, which took him to 1.3 centimeters (half an inch) above the schoolroom globe. Mike was nonetheless far enough away for Earth to reveal itself through the lens of the universe and not the lens of geopolitics—the overview effect in full force. How many on Earth, among those who are oppressed, or at war with their neighbors, or who hunger for food, would possibly think of Earth, itself, as Heaven? Gaining that on-orbit perspective and returning to Earth changes your relationship with our planet and with your fellow humans.

Turns out, two regions of the world exist where you can indeed identify a national border from space, as if mapmakers drew it themselves. In one area, fields of green, and on the other side of a sharp boundary, brown desert. In another area, the nighttime landscape is ablaze with city lights, yet across a sharp boundary, the depths of darkness. This can happen naturally at a mountain range or some other extensive natural barrier. Not here. These boundaries are sharp and thin. They reveal unevenly shared resources across borders, where one side controls their landscape while the other side does not.

What's going on?

The irrigated/desert borders are in the Middle East, between Israel and the Gaza Strip, and between Israel and large swaths of the West Bank—regions of ceaseless political tension. The GDP per capita of Israel is twelve times larger than that of Palestine.[2]

The light/dark border is in East Asia, between South Korea (Republic of Korea) and North Korea (Democratic People's Republic of Korea). The GDP per capita of South Korea is twenty-five times larger than that of North Korea. Another region of persistent political tension.

Such stark visual differences from above need not be national boundaries of disparate geopolitics. They can also delineate areas of subjugation. In 1992, during my first of several visits to South Africa—two years before Nelson Mandela was elected president—I flew into Johannesburg at night and noticed on the long approach to the city a large ground area with sharp boundaries and no interior lights. It was clearly a lake. That's what lakes look like from an airplane at night. Returning home a week later, flying back out in the daytime with the ground fully illuminated, what I had seen was no lake. It was the larger part of Soweto, an all-Black shantytown of Johannesburg with no electricity. No lighting at night, or at any other time of day. No appliances. No refrigerators. I can know all this intellectually for having read about it, but when I confront it from above, with details stripped away, I don't see the politics, the history, the skin colors, the languages, the bigotry, the racism, the protests. I'm instead plagued by a simpler thought that, as a species, we might not possess the maturity or wisdom the future requires to assure the survival of civilization.

Let's ascend some more. In fact, let's go all the way to the Moon. Here's a quip shared occasionally among space enthusiasts:

If God wanted us to have a space program,
he would have given Earth a Moon.

An invertedly insightful comment on what drives space exploration. Note that Venus, our nearest planetary neighbor, has no moon at all. Mars, our next-nearest neighbor, has Phobos and Deimos, two sorry excuses for moons. Both are Idaho potato-shaped and tiny enough to fit easily within the municipal boundaries of most cities. Our Moon, on the other hand, is 50 percent larger than Pluto and checks in as the fifth largest in the Solar System. So Earth did well in the moon lottery—endowed with a destination of dreams for intrepid explorers.

If the Space Shuttle and the International Space Station orbit between 1 and 1.3 centimeters above the schoolroom globe, find the Moon in the next classroom, ten meters away. That's why it takes only eight minutes for a rocket to reach Earth orbit, but a full three days for the Apollo rockets to reach the Moon.

The seminal photo of planet Earth rising over the lunar landscape—you know the one—arrived in December 1968 from *Apollo 8*, the first mission ever to leave home for another destination. From deep space the entire Earth is laid bare. As the cosmos intends you to see it: a fragile juxtaposition of land, oceans, and clouds; isolated and adrift in the void of space, with no hint of anybody or anything coming to rescue us from ourselves. This outlook ascends a few realms higher than the overview effect and represents the true beginnings of a cosmic perspective.

Earth in the lunar sky is nearly 14 times larger than the Moon in Earth's sky. Earth is also about 2.5 times more reflective than the Moon, although the exact value from moment to moment varies with cloud cover. So full Earth viewed from the Moon is about 35 times brighter than full Moon viewed from Earth. Unlike what's implied by NASA's Earthrise photo, snapped from the orbiting *Apollo 8* command module, Earth neither rises nor sets on the Moon. Viewed from the Moon's near side, Earth never leaves the sky. From the Moon's far side, you might never know Earth existed at all.

Apollo 8 did not land on the Moon, which left them forgotten in the shadow of *Apollo 11*'s Neil Armstrong, Buzz Aldrin, and Michael Collins. Instead, *Apollo 8* "just" orbited the Moon ten times before returning. Yet no one had seen Earth so distant before. Their mission from launch to splashdown lasted from December 21 to December 27, which closed out the bloodiest year of the most turbulent decade in the US since the Civil War, a century earlier. During the Vietnam War, more Americans and Vietnamese died in 1968 than in any other year. A year that included the Tet Offensive in February and the infamous Mỹ Lai Massacre in March. Back home now add the assassinations of Martin Luther King Jr. in April and Bobby Kennedy in June, followed by persistent violent protests in cities and on college campuses.

Some say *Apollo 8* saved 1968.[3] I prefer to say *Apollo 8* salvaged it.

Notice that *Apollo 8*'s journey enclosed the Christmas holiday. On Christmas Eve, while still in orbit around the Moon, the three astronauts—Bill Anders, James Lovell, and Frank Borman—took turns reading the first ten verses of the Book of Genesis from the King James Version of the Bible. You may be familiar with some, if not all of these lines.

Anders began . . .

We are now approaching lunar sunrise, and for all the people back on Earth, the crew of Apollo 8 *has a message that we would like to send to you.*

> *In the beginning God created the heaven and the earth.*
>
> *And the earth was without form, and void; and darkness was upon the face of the deep. And the Spirit of God moved upon the face of the waters.*
>
> *And God said, Let there be light: and there was light.*
>
> *And God saw the light, that it was good: and God divided the light from the darkness.*

Lovell continued . . .

> *And God called the light Day, and the darkness he called Night. And the evening and the morning were the first day.*
>
> *And God said, Let there be a firmament in the midst of the waters, and let it divide the waters from the waters.*
>
> *And God made the firmament, and divided the*

WATERS WHICH WERE UNDER THE FIRMAMENT FROM THE WATERS WHICH WERE ABOVE THE FIRMAMENT: AND IT WAS SO.

AND GOD CALLED THE FIRMAMENT HEAVEN. AND THE EVENING AND THE MORNING WERE THE SECOND DAY.

Borman finished . . .

AND GOD SAID, LET THE WATERS UNDER THE HEAVEN BE GATHERED TOGETHER UNTO ONE PLACE, AND LET THE DRY LAND APPEAR: AND IT WAS SO.

AND GOD CALLED THE DRY LAND EARTH; AND THE GATHERING TOGETHER OF THE WATERS CALLED THE SEAS: AND GOD SAW THAT IT WAS GOOD.

And from the crew of Apollo 8, *we close with good night, good luck, a Merry Christmas—and God bless all of you, all of you on the good Earth.*

The cosmic perspective can do that to you, especially if you are religious. Back on Earth, in response to this biblical reading beamed from the Moon, noted atheist Madalyn Murray O'Hair, founder of the ardent organization American Atheists, sued the federal government, accusing it of violating the First Amendment—the part that says "Congress shall make no law respecting an establishment of religion." The suit was dismissed from all levels of court to which it was taken. With no legal expertise, I cannot comment on the legitimacy of her lawsuit, but I have thoughts on the existence of the lawsuit itself. At the time, I was only ten years old, but had the me-of-today met the Madalyn-of-then, here's how that conversation would have gone down:

Neil: Were you strapped to a *Saturn V* rocket, with nine million pounds of thrust, and sent a quarter-million miles away into deep space to witness Earth-rise from lunar orbit on Christmas Eve?

Madalyn: No.

Neil: Then shut the fuck up.

In that first of nine such missions, our goal was to explore the Moon, but while doing so, we looked back over our shoulders and discovered Earth for the first time. After the Earthrise photo made its rounds, people changed. Earth-the-planet mattered as never before. If your local river or lake got polluted by factory effluences, you would get angry and try to do something about it. But thinking about all of Earth as a holistic ecosystem, and not just your section of it, had not yet become a priority, or even a thought.

In the US, important seeds had been planted as early as 1872, with the congressional designation of Wyoming's Yellowstone as the first national park, followed by President Teddy Roosevelt's 1906 Antiquities Act to preserve natural and otherwise historic monuments, and by President Woodrow Wilson's National Park Service in 1916. Now fast-forward to the 1962 publication of marine biologist Rachel Carson's best-selling *Silent Spring*, itself a wake-up call to the consequences of rampant pesticide use in agriculture, especially the environmental impact of dichloro-diphenyl-trichloro-ethane, commonly (and sensibly) abbreviated as DDT. The book's content and success prompted President John Kennedy to ask his science advisory committee to study the problem. And in early 1969, the US experienced a nasty oil spill, with up to 16,000 cubic meters of crude oil polluting the waters and beaches of wealthy Santa Barbara County[4] in California.

By 1970 we are poised and positioned to save the planet.

The full spate of nine Apollo lunar missions spanned December 1968 through December 1972. Over that time, we were still fighting the Cold War with the Soviet Union and a hot war in Vietnam. Campus unrest continued, culminating in the 1970 Kent State University shootings in which unarmed anti-war protesters were gunned down by the Ohio National Guard, who wounded nine and killed four.[5] The United States clearly had pressing problems to solve. Yet we paused to reflect on our relationship to Earth.

What follows is a list of actions, unprecedented in their rapidity, their scope, and their mission, all occurring between 1968 and 1973, well before corresponding legislation would be passed in other industrialized nations of the world.

Apollo 11 (First to walk on the Moon)	1969
Comprehensive Clean Air Act	1970 (Earlier versions: 1963, 1967)
First National Earth Day	1970
National Oceanic and Atmospheric Administration (NOAA) formed	1970

Whole Earth Catalog published	1968–1972 (Sporadically thereafter)
Environmental Protection Agency (EPA) formed	1970
"Crying Indian" public service TV commercial	1971
Doctors Without Borders founded, Paris	1971
DDT banned	1972
Clean Water Act	1972
Apollo 17 (Last mission to Moon)	1972
Endangered Species Act	1973 (Earlier versions: 1966, 1969)
First catalytic converter for cars	1973
First unleaded gasoline emission standards	1973

A few notes: The "Crying Indian" was an emotional appeal to not throw trash out your car window and became one of the most recognized public service announcements of the era, even though the actor dressed in Native American garb while shedding a tear was actually of Italian descent. I include the founding of the humanitarian organization Doctors Without Borders on the supposition that without a view of Earth from space, the very concept of "without borders" might not have arisen, instead perhaps referencing "international" doctors, or doctors "crossing borders." Just a thought. Also, the official Earth Day history page[6] makes no mention of Apollo or images of Earth from space, simply declaring that Rachel Carson's 1962 book was what primed the pumps for the 1970 inaugural celebration of Earth. Meanwhile, images of Earth from space became the endearing and enduring symbol of the *Whole Earth Catalog* and an unofficial flag for Earth Day,[7] displaying *Apollo 17*'s image of full Earth taken on their return from the Moon in December 1972.

One can never know for sure the subtle or not-so-subtle causes and effects of human actions. In my understandably biased view, the I-care-about-Earth movement could have, in principle, happened in 1950 or 1960. The environment was no less a problem then. For example, air pollution in the Los Angeles basin, from the rise of cars combined with unfortunate thermal inversions due to local geography, made LA during the 1940s and 1950s one of the most polluted cities on Earth.[8] Or it could have happened while Rachel Carson's book was still on the *New York Times* Best Seller list, where it lived for thirty-one weeks from 1962 to 1963. That would have been ideal, but Kennedy was assassinated in 1963, and Carson sadly died of cancer in 1964. Between 1963 and 1969, a total of four governmental reports spanning three presidents would be filed, each exploring the effect of pesticides on crops and human health, and each calling for the phaseout of DDT.[9] Still no legislative action over that time. Before the 1969 Santa Barbara oil spill, two others occurred, from December 1962 to January 1963, affecting the Mississippi and Minnesota riverbanks.[10] Still, we were all distracted by other things. Mass environmental awareness could have waited until 1975, after the end of the Vietnam War. Or until 1990, after the fall of the Berlin Wall. No. It all happened smack in the middle of our Apollo missions to the Moon—while a starry message was gently descending from space, infusing us all with a veritable firmware upgrade of our collective capacity to care.

Apart from serving as our first destination in space, the Moon enjoys extraordinary value across world cultures. The cycle of lunar phases informs the reckoning of time for the Chinese, Islamic, and Hebrew calendars. Also, full moons rise at sunset and set at sunrise, and the physics of reflectivity makes them six times (!) brighter than half-moons, so they provide an excellent guiding light throughout the night. The Moon also feeds our superstitions, believed by some to exert mysterious powers of influence on human behavior, especially during full Moon.

The art of asking questions represents an important dimension of science literacy. How you think and how you query nature matter more than what you know. Often the answers reveal themselves simply by asking the right questions in the right sequence.

Everyone knows the full Moon turns some of us into werewolves, but how come it doesn't happen when you're in the basement, or if there's complete overcast? Just because you can't see the full Moon through the clouds doesn't mean the Moon isn't there. So it must be the light that temporarily converts your genetic profile to that of a wild canine. Holding aside the biological and physiological implausibility of this claim, did you know that moonlight is just reflected sunlight? Obtain a spectrum of the Moon's light and it's identical to that of the Sun. (A fact I demonstrated for my eighth-grade science fair project using a spectroscope I had built from scratch. Came in second place.) If the Moon turns you into a werewolf at night, then so should the Sun in the daytime.

Some hold the opinion that more babies are born under the full Moon than any other phase. It so happens there's no statistical evidence to support this claim, only anecdotal, and the full Moon exerts no extra gravity on Earth (or on us) relative to its other phases. Let's pretend we're unaware of those salient facts. For now, we're just asking questions. Was the delivery table facing the full Moon in the sky when the baby was born? If so, maybe the Moon helped yank the baby from the womb. Presumably, delivery tables orient randomly across all hospital delivery rooms. We then expect some of these tables to face opposite the full Moon during delivery. In those cases, any extra Moon gravity would now pull the opposite direction, keeping the baby in its mother's womb and delaying the birth, not speeding it up.

If not gravity, then perhaps some mysterious other force was operating. How about something the methods and tools of science have yet to discover. New science is always fun, and typically leads to Nobel Prizes all around, but let's not get ahead of ourselves. We're still just asking questions. How long is the human gestation period? Doctors will tell you 280 days (40 weeks), which is not entirely true. That's the count of days from the last menstrual cycle. Hardly anybody gets pregnant then. You probably got pregnant when you ovulated, two weeks later. So the actual time to make a full-term human baby comes to 280 days – 14 days = 266 days. Now let's ask about the time it takes for the Moon to cycle through its phases—full Moon to full Moon. You can look that up: it averages 29.53 days. Nine of these cycles comes to 266 days.

That's interesting. A full-term baby takes about nine cycles of the Moon to develop in the womb. So to be born under a full Moon, romantic as it is, simply means you were likely conceived under one, romantic as it was. No need to appeal to new physics or mysterious forces or supernatural events to explain the non-result. Rational inquiry took you there.

The Moon affects the waters of Earth, with especially high oceanic tides during full Moon. Biologists tell us we are mostly water, as are the oceans. Surely then, the full Moon's tidal forces affect us in some way, if not to turn us into lunatics. In the Earth-Moon system, the side of Earth facing the Moon is closer, so it feels a slightly stronger force of gravity than the side facing away. This creates a stretching force across Earth's diameter, most visible in our oceanic tides, but the solid earth experiences it too. Notice that this description of Moon tides on Earth makes no mention of the Moon's phase. That's because the strength of the Moon's tides has nothing to do with the Moon's phase. Full Moons get the highest tides not because of the Moon but because of the Sun. The Sun's tides on Earth are about one-third the strength of the Moon's tides, yet hardly anybody talks about them. During full Moon, high Sun tides add directly to high Moon tides, giving the false impression that the full Moon imparts extra gravitational influence. Furthermore, whatever false gravitational influence you ascribe to the full Moon, you must also grant to the new Moon because the Sun's tides perfectly align there too.

Another way to think about it: How does the Moon's tidal force across the diameter of your head compare with the Moon's tidal force across the diameter of Earth? Like Earth, at any moment, there is one side of your head closest to the

Moon, feeling stronger gravity than on the other side. Now imagine that your noggin was just a seven-inch sphere of ocean water. How much distortion can we expect from the Moon's tidal forces? When you do the math, you get about one-ten-thousandth of a millimeter. That's substantially less than the distortion on your skull from the weight of your own ten-pound head on itself, as you sleep on pillows at night. Yet nobody writes lycanthropic stories about what brand of pillow you use, or how wide your head might be. If that bedtime example is too abstract, just imagine sticking your head in a vise and having your friend tighten it to a force of ten pounds—every night.

One of the rewards for doing well in the moon lottery is that the Sun is four hundred times wider than the Moon, and it happens to be four hundred times farther away. This pure coincidence renders the Sun and Moon about the same size in the sky, allowing for spectacular solar eclipses. This wasn't always the case, nor will it be so in the distant future. The Moon is spiraling away from Earth at a rate of about 1.5 inches per year. So let's enjoy this match made in heaven while we can. Every few years, the Moon passes exactly between Earth and the Sun, precisely covering the Sun, darkening the sky, and briefly laying bare the Sun's gorgeous outer atmosphere, called the corona. No other planet-moon combination in the Solar System can match it.

Eclipses lead a long list of sky phenomena that irresistibly attract and entangle us. The idea that the Sun, Moon, planets, and stars affect us personally is called astrology and goes way back. Some call it the second-oldest profession. How could any of us think any different as we watched the sky revolve around us daily. For example, certain constellations rise before dawn every autumn, just when your crops are ready for harvest. Clear evidence that the entire dome of the sky, day and night, lovingly looks after your needs and wants.

With this mindset, the sky also portends events you, your culture, or your religion might either desire or dread. Some years ago, I received a phone call at my Hayden Planetarium office from a woman who was rummaging through her deceased father's personal effects, which included a diary from one of her distant ancestors—a settler in New England who described his family's migration via covered wagon to a new home. The diary fretfully describes a rapid darkening of the skies in midday. The patriarch of the family was deeply concerned that, without warning, the sixth seal had been opened, and a biblical prophecy from the Book of Revelation was underway:[11]

I looked when He opened the sixth seal . . .
and the sun became black as sackcloth of hair,
and the moon became like blood.

He stopped the wagon and prompted his entire family to kneel, pray, repent, and ready themselves to meet their maker. The caller to my office wondered if her ancestor had unknowingly described a total solar eclipse. I asked for the date of the diary entry. She wasn't sure, but knew the trip was in the early 1800s. From specialized software I keep just for emergencies such as this, I localized the moment to midday, June 16, 1806, and affirmed they were in Massachusetts. In the early nineteenth century, total solar eclipses were fully understood by any mildly educated person who witnessed them. So the only way the event could be misinterpreted is if the wagon's location was under total cloud cover. Now combine an inexplicably darkened sky with a deeply religious, Bible-fluent Christian, and you are pleading for God's mercy. Without the clouds, the solar eclipse becomes an educational spectacle for the whole family.

Twenty-seven years later, in the predawn hours of November 13, 1833, the annual Leonid meteor shower was particularly memorable. Visible across all of North America, it delivered upwards of one hundred thousand "falling stars" per hour. That's as many as thirty per second. For comparison, a good meteor shower might average one per minute, so this was a shocking display. We call meteor showers this intense "meteor storms," which result from Earth plowing through a denser-than-normal pocket of cometary debris as we orbit the Sun. The bits of comet Tempel-Tuttle that comprise the Leonids are no larger than peas. But on colliding with Earth's atmosphere at 160,000 mph, they burn up as streaks of light, visible in darkened skies. At the time, future president Abraham Lincoln, then twenty-four years old, lived in Illinois as a boarder with a deacon of the local Presbyterian church. Upon witnessing this unforgettable cosmic display, the deacon swiftly aroused Lincoln, declaring, "Arise, Abraham, the day of judgment has come!" A conclusion drawn from yet another—the very next—apocalyptic verse in the Book of Revelation:[12]

And the stars of heaven fell unto the earth,
even as a fig tree casteth her untimely figs,
when she is shaken of a mighty wind.

Honest Abe, a self-educated, lifelong learner, dutifully went outside to look up at the night sky. Invoking his knowledge of astronomy, he noticed that all the grand constellations were still there, intact—Ursa Major, Leo, Taurus, Orion. Whatever was falling, it wasn't the stars, and so he rationally concluded that a biblical prophecy of doom was not underway,[13] and promptly went back to bed.

Knowledge of the universe—cosmic perspectives in particular—simultaneously disconnects our ego from all that happens in the sky yet fosters accountability for all we do, preventing us from either crediting or blaming the sky for our earthly affairs. In other words, from Shakespeare's *Julius Caesar* (1599):

The fault, dear Brutus, is not in our stars
But in ourselves.

In 1994, Carl Sagan published *Pale Blue Dot: A Vision of the Human Future in Space,*[14] inspired by a 1990 photo of Earth taken by the *Voyager 1* space probe after crossing the orbit of Neptune. That moment was chosen as an external visitor's first encounter with our Solar System's planets, coming the other way. Earth occupies barely a pixel in that image, offering a stark reminder of how small we are in the universe. In his book, Carl waxes poetic about how frail and precious Earth appears when seen as a pale blue dot. That image comes closest to a next-generation reset of our societal mission statement, even if it falls shy of having the impact of the Earthrise photo. In any case, one might think of it as Earth's first selfie.

Other cosmic revelations that could rival Earthrise include the discovery that we are not alone in the universe. This could signal a change in the human condition that we cannot foresee or imagine. Taken to enticing yet frightening limits, we might exist in a computer simulation programmed by intelligent juvenile aliens still living in their parents' basement. Or we might discover that planet Earth is a zoo—a literal terrarium/aquarium—constructed for the amusement of alien anthropologists. Going even further, perhaps our cosmos, complete with our hundred billion stars per galaxy and the hundred billion galaxies in the observable universe, is nothing more than a snow globe on some alien creature's mantel.

In all these scenarios, the cosmic perspective will have morphed from the universe reminding us to take better care of our own fate to the universe declaring we're playthings for high-level life forms. A terrifying prospect, perhaps. But we take better care of our cats and dogs than we do of homeless humans in the street. If we serve as pets to aliens, might they take better care of us than we ever will of ourselves?

FOUR

CONFLICT & RESOLUTION

Tribal forces within us all

One of the great features of a working democracy is that we get to disagree without killing one another. What happens when democracy fails? What happens when we hold no tolerance for views that differ from our own?[1] Do we instead desire a dictatorship in which all views of the land agree with those of the dictator? Do we pine for a system where dissenting views are suppressed, buried, or burned? Do we long for a world where the moral code, our values, and our judgments—all that we believe is right and wrong—are deemed correct and unassailable?

Part the curtains of all-out conflict and find the puppeteers of politics and religion. Two topics that, we are warned, should never be discussed in polite company. Two topics with a lot in common for how deeply personal they can be. Two topics that, when disagreements are severe, can lead to bloodshed and full-blown war.

Including all casualties among all warring nations across the six-year span of World War II (1939 through 1945), more than one thousand people were killed—per hour. A morbid and inevitable consequence of forcing your personal truths upon others in a world that is fundamentally pluralistic. A scientist's entire mission in life is to discover features of nature that are true, even if they conflict with your philosophies. That's why you'll never see battalions of astrophysicists storming a hill. No doubt, scientists and their military handiwork have been the pawns of military ideologues[2] from the very beginning of things. Most scientists, however, have neither motive nor agency to do this all by themselves—on their own volition. Even Wernher von Braun, the architect of Apollo's rockets to the Moon, famously commented on the success of the V2 ballistic missile,[3] which he pioneered for Nazi Germany and launched primarily against London and Antwerp,[4]

The rocket worked perfectly, except for
landing on the wrong planet.

The in-group/out-group behavior of human beings throughout civilization is particularly disturbing, even if evolutionarily understandable.[5] If we can't fully overcome our DNA, maybe an infusion of evidence-based thinking can penetrate the no-evidence posturing. Consider what happens when scientists disagree. We look for one of three outcomes: either I'm right and you're wrong, you're right and I'm wrong, or we're both wrong. That's an implicit contract we carry into all arguments on the frontier of discovery. Who decides the outcome? Nobody does. Arguing more loudly or strenuously or more articulately than your opponent simply reveals how annoying and obstinate you are. The resolution almost always comes when more or better data arrive.

In rare cases, both sides of an argument can be right, but only when they unknowingly describe disparate features of the same object or phenomenon—like the proverbial blind men who each describe their encounter with an elephant—the tusk, the tail, the ears, the legs, the trunk. They could argue all day about who is right and who is wrong. Or they could keep exploring, and eventually figure out that these are distinct parts of a single animal. That too requires more experiments and observations—more data—to determine what is objectively true.

Apart from conflicts over politics or over what God or gods you may worship, humans persistently wage war over access to limited resources, such as energy (oil and gas), clean water, mineral deposits, and precious metals. In our cosmic backyard, solar energy is ubiquitous, as are freshwater comets. Metallic asteroids are there too, calmly orbiting the Sun in abundance. The big ones each contain more gold and rare-earth metals than have ever been mined in the history of the world. We're not there yet, but imagine the day when all of civilization becomes spacefaring. Routine access to space will turn the Solar System into Earth's backyard. With that access comes unlimited space-borne resources, rendering an entire category of human conflict obsolete. Access to space may be more than just the next frontier to explore; it could be civilization's best hope for survival.

Of all professions, scientists may be uniquely capable

of generating and sustaining peace among nations. We all speak the same basic language. For example, the mathematical value of pi does not change as you cross passport control at a nation's border. The laws of biology, chemistry, and physics remain the same too. And we share a common mission statement—to explore the natural world and decode the operations of nature along the way. Here's how that might play out: Imagine you're on a space mission at a Moon outpost, collaborating on science experiments with a fellow astronaut who's a different nationality from your own. Back on Earth, for whatever reasons, geopolitical tensions escalate between your two countries. Relations get so bad that both countries withdraw their ambassadors from each other's embassies. Armed conflict results, leading to mass casualties of soldiers and civilians. You're in space—you're on the Moon—what do you do? Do you wrestle your fellow space collaborator to the ground, consumed by the emotions, angers, and actions of Earth politicians 238,000 miles away? Or maybe your heads of state simply radioed you both with instructions to break off all contact with each other. Would you? Should you? Or do you peaceably go on about your Moon-day, performing experiments, even if dismayed and ashamed that you are both members of a species that for thousands of years has made an art and a science of killing one another on Earth.

Not all countries get to explore. Those that do share a bond that rises above all that may divide us. On government assignment, while serving on the first of my two White House commissions, this one on the "Future of the

US Aerospace Industry,"[6] I had the occasion to meet and greet counterparts of mine in many countries as we explored the world's landscape of aerospace as a provider of transportation, commerce, and security. Across Europe, Russia, and East Asia, we assessed the challenges and opportunities that may await us in our future. Throughout, I enjoyed excellent camaraderie with fellow scientists and engineers. The politicians and business executives on this commission also felt warmly received. But the atmosphere was curiously elevated with the Russian representatives. At Star City, the training center for cosmonauts just outside Moscow, the vibe was transcendent. I don't speak Russian. I can't pronounce their words because I don't know the Cyrillic alphabet. And I don't enjoy vodka, which was served to us shortly after our arrival—around 10:00 a.m.—by the head of the facility, from a secret door behind his desk. After they invited *Apollo 11* moonwalker and fellow commissioner Buzz Aldrin to sign their big book of spacefarers, we all started talking about the space race of the 1960s and 1970s, and the future of space exploration. That's when all barriers evaporated, and I felt like I had known every Russian in the room my entire life—as though we were childhood friends who had played together with the same toys in the same sandbox.

The United States and Russia (USSR) were the only countries with the capacity to put a human into orbit for forty-two years, until China joined the club and launched their first taikonaut in 2003. Between and among us, the emotional connections were deep, and the friendships rose far above earthly politics. We shared a bond forged in space.

I had been raised during the Cold War and, like other red-blooded Americans, saw the Russians as evil, godless, commies. Didn't we . . . hate each other? Weren't we . . .

mortal enemies? Actually, no. The politicians were all that. Instead, our sights were on the stars the whole time—as explorers—empowering a world perspective that transcends the conflicts of nations every time.

The top two most costly displays of international cooperation, in order, are the waging of war and the construction and operation of the International Space Station. Others include the Olympics and the World Cup. Three of those four engage in competition, one of which causes the loss of human life. As for the Space Station, the list of countries that sent astronauts there includes Belgium, Brazil, Denmark, Great Britain, Kazakhstan, Malaysia, the Netherlands, South Africa, South Korea, Spain, Sweden, and the United Arab Emirates, not to mention the United States, Russia, Japan, Canada, Italy, France, and Germany. That's fewer flags than fly over the World Cup or the Olympics, but a brief review of twentieth-century geopolitics, still etched in memory of people living today, reminds us how many of those same countries fought each other in all-out war, with soldier and civilian death tolls in the millions.

During the early 1970s, the US and the USSR were still holding the world hostage at thermonuclear gunpoint, and the Cold War would not end for another two decades. Meanwhile, in 1972, President Richard Nixon and Soviet premier Alexei Kosygin—in Moscow—signed an agreement to launch the Apollo-Soyuz Test Project. Three years later, in July 1975, astronauts and cosmonauts executed a first-ever space rendezvous in a docking maneuver between our Apollo command module and their Soyuz capsule. The only rule when they popped the hatch? Americans would speak only Russian, and the Russians would speak only English.[7]

As we say to students who may be struggling with Astro

101, "The universe is above everyone's head." So too may be the most fertile prospects for world peace.

I lean liberal in nearly all relevant opinions that I hold. Yet it was President George W. Bush who twice appointed me to serve on White House commissions. He sought my scientific expertise, and my politics didn't seem to matter. My opinions are my own, and (believe it or not) I invest relatively little effort to get people to agree with them, so perhaps my mostly muted politics softened any latent concerns he may have had by crossing the aisle to include me.

That appointment was a political baptism for me. I met and became friends with staunch conservative policy wonks as well as progressive labor leaders. In this politically diverse and powerful group of twelve commissioners, successful conversations were the ones that took place in the middle of the political spectrum. This meant I had to migrate from my left-leaning corner and bring my perspectives closer to those who I persistently disagree with. My steps were tentative but refreshing. Every inch closer to the conservative worldview I stepped took me that much farther from the liberal worldview I had known. This continued until the moment I realized, for the first time in my life, that I was truly thinking for myself—no longer torqued by ideologies I was born into and had adopted as my own without question. I saw conservatives for the first time as something other than a monolith. I also saw liberals for the first time—aided and enabled by this unfamiliar but luminous view from the center. While there, I came to resent labels of all kinds. What are they, if not intellectually lazy ways of asserting you know everything about a person you've never met?

Can scientific rationality embedded in a cosmic perspective

make everyone agree on all opinions? No, not likely. Although, it can make everyone disagree less strenuously—not from compromise, but from the inevitable separation of your emotions from your capacity to reason, and the reduced bias in your capacity to think. Sometimes, all you need is more or better data.

Consider four red-blue political tropes in the US, and what they look like to an inquisitive scientist:

> TROPE 1: Conservatives value the nuclear family and the stability it brings to civilization, unlike liberals, who live under questionable moral codes.

Let's look at family values by the usual issues that come under scrutiny: out-of-wedlock births and divorce rates. If you analyze state-by-state childbirth statistics, you find that nearly half of all babies born in the states of Louisiana, Alabama, Mississippi, Texas, Oklahoma, Arkansas, Tennessee, Kentucky, West Virginia, and South Carolina are born to unmarried women.[8] Yet each of these states voted red in every general election this century.[9] The corresponding rates for the famously blue states of California, Minnesota, Massachusetts, and New York are half that. Babies born out of wedlock could highlight, for example, a landscape of liberated women who don't need men or eschew 1950s family paradigms. Or it could indicate regional differences in abortion rates. In any case, it's not evidence of traditional family values.

How about divorce rates across the country? States with the lowest rate might indicate a culture of stable family life. When you rank all fifty states (for the year 2019), you find that six of the top ten states with the lowest divorce rates are blue. Four are red. Okay. No case to be made there. But let's look a little closer: The states with the two lowest rates, by far, are Illinois and Massachusetts, both persistently blue. Additionally, nine of the ten states with the highest divorce rates voted red in the 2020 election. Also, the only two sitting US presidents with a previous divorce on their record were Republicans Ronald Reagan and Donald Trump. President Trump has divorced twice. Melania is his third wife. His second wife was who he cheated with on his first wife. And he has children with all three women.

What does one do with these facts? Sweep them under the rug? Or have them defuse any emotionally charged arguments about which political party carries the standard of morality?

But wait. In 2015 a data breach of the infamous Ashley Madison online dating service produced an accidental statistic to further illuminate this landscape of misperceptions. The website is conceived for a married person to hook up with another married person who both want to cheat on their spouses. When revealed which states were most active on the site, the left-leaning states such as New York, New Jersey, Connecticut, Massachusetts, Illinois, Washington, and California landed in the top fifteen among the cheaters.[10] So maybe the truth is more subtle and complex than either the Right or the Left would ever admit. Maybe divorce is a more honest resolution of a failed relationship than seeking secret, illicit affairs while staying married. In either case, we learn from rational inquiry that no part of the political spectrum may lay claim to morally superior family values.

TROPE 2: Liberals occupy the high ground on science while conservatives embrace science deniers.

Where do I begin? The denial of climate change poses an existential threat to the stability of civilization—a position that lands squarely amid conservative platforms, although there's been some progress over the years. Initially the conservative battle cry was denial, which later morphed into an admission that climate change is real, but humans aren't causing it. For some, that ultimately became an admission that humans are causing it, but there's nothing we can or should do about it. Here is an actual sentence from the 2018 official Republican Platform[11] of the oil-dependent[12] state of Texas:

Climate change is a political agenda promoted
to control every aspect of our lives.

A classic case of wishful thinking in which political convictions override objective truths. Updated every two years, by 2020, that line was removed, leaving behind only:[13]

We support the defunding of "climate justice" initiatives.

We've all seen the numbers. Upwards of 97 percent of climate scientists agree[14] that our industrialized civilization, sculpted from highly transportable, energy-dense fossil fuels, is boosting Earth's greenhouse effect, which is melting glacial ice and will eventually flood all coastal cities of the world. This conclusion comes from more than a majority vote. It derives from a body of research bolstered by repeated observations and experiments across multiple disciplines, which is just what you need and want before you declare a new objective truth in the world. Denial here is to live in the 3 percent of the research papers that conflict with or flatly deny the prevailing results.

To help bring perspective to the scientific consensus, let's invoke a "thought experiment." These are time-tested tactics used by many scientists, famously by Albert Einstein, as a thinking person's way to imagine an experiment that you don't have the time or money to conduct. How about: A bridge is on the brink of collapse and 97 percent of the structural engineers tell you, "The bridge will fall if you drive your truck across it. Take the tunnel instead." The remaining 3 percent said, "Don't listen to them; the bridge is fine!" What would you do? Or how about: An untested suicide pill is invented that 97 percent of medical professionals say will kill you with one dose. Although 3 percent say you'll be just fine, and it may even improve your health. If you wanted to improve your health, would you take the pill? Thought experiments, when conceived to shift the context of a question in time, space, and scope, can ferret out hidden bias, coercing you to confront your own foundations of thought, possibly for the first time.

Right-leaning rebuttals to climate change have continued to evolve. The most recent variety accepts the premise that humans are warming the planet, but argues strenuously with left-leaning people about the economics of it all. In particular, they worry that platforms such as the Green New Deal[15] will trigger financial catastrophe. Hooray! We've finally arrived: a politically charged conversation about what policies should be enacted in response to scientific truths. That's how an informed democracy is supposed to work.

In another hotbed of science denial, some conservative Christians doubt Darwinian evolution because their 3,500-year-old sacred texts lay out a different idea for how all the animals and the rest of life on Earth came to be. These fundamentalists are simply exhibiting their constitutionally protected free expression of religion. They happen to be the minority of Christians,[16] and I'm not interested in changing their views unless they try to overthrow the country's science curriculum or lobby to become head of a government science agency. Plenty of high (and low) paying jobs exist that don't require you to accept the tenets of modern biology.

All that's conservative is not anti-evolution. Consider the landmark 2005 court case *Kitzmiller v. Dover Area School District* of Pennsylvania, in which federal judge John E. Jones III ruled that the teaching of God-inspired "intelligent design" in public schools was unconstitutional. Judge Jones was appointed to that post by Republican president George W. Bush.

Other than climate change and modern biology, there's very little else about science in America that conservatives deny, despite liberals lobbing broad science-denying accusations at them. Okay. Then what about the liberals themselves? Turns out, the following list of beliefs and practices lands squarely in their corral: crystal healing, therapeutic touch, feather energy, magnetic therapy, homeopathy, astrology, anti-GMO, and anti-pharma. What all these ideas and movements have in common is a flat-out rejection of some or all mainstream science relevant to each subject. Before the Trump administration and the 2020 conservative-fed resistance to the rapidly developed COVID-19 vaccine, the anti-vaccine movement (another science-rejecting platform), was led primarily by liberal-leaning communities. They practically invented the brand. For example, in 2000 the World Health Organization declared measles in the US to be "eliminated" based on the success of ongoing vaccine programs.[17] Yet in 2019 the US logged nearly 1,300 cases of it. Who hosted most of those outbreaks? The perennially blue states of Washington, Oregon, California, New York, and New Jersey, where many parents refuse[18] to get their children vaccinated. Now, with the anti-vax movement turning purple—as it spills into conservative enclaves[19]—the combined total of anti-vaxers may represent as much as one-fourth of the nation.[20]

In an August 2021 posting that, in retrospect, should have stayed in my "Forbidden Twitter" file, I looked at the number of American citizens dying per day of COVID-19's Delta variant—around 1,000. I noticed at the time that at least 98 percent of everyone who was hospitalized and dying from COVID-19 were unvaccinated. From various surveys, I saw that five times as many people who vote Republican remain unvaccinated relative to those who vote Democrat. If you do the math, you arrive at my post:

> *Right now in the USA, every ten days, more than 8,000 (unvaccinated) Republican voters are dying of COVID-19. That's 5X the rate for Democrats.*

To that, I affixed a meme that showed a mock book title in Gothic font:

How to Die Like a Medieval Peasant
in Spite of Modern Science

Within seconds, all manner of Twitter fights broke out. Many conservative anti-vaxers stood their ground and doubled down on their decision to remain unvaxed—for freedom's sake. Others chose to unfollow, accusing me of politicizing COVID. Some questioned the source data. Still others complained that I shouldn't make light of people's deaths. Even my "woke" daughter called to tell me it was harsh. I foresaw none of those reactions, thinking instead that people, especially Republicans, would say, "Hmm. That's bad. We need more, not fewer, voters in the midterm elections. Let's get vaccinated." When I misfire that way, it means I failed as an educator to understand and navigate people's receptors for absorbing my post. I deleted the Tweet and replaced it with the link to one of my podcasts on vaccine science[21] with a medical professional.

These vax facts among conservatives notwithstanding, most of the liberal-leaning beliefs will not precipitate the end of civilization. Liberal science denial as currently expressed will never destabilize the world as much as the conservative science denial of climate change. So liberals today can claim their activities are better for the planet, but they cannot smugly brand themselves as pro-science.

In recent years promoters of questionable dietary supplements have infiltrated the sponsors of radical, right-wing radio shows and podcasts. Brain Force Plus, Super Male Vitality, Alpha Power, DNA Force Plus, are all examples of non–FDA approved pills and extracts offered for sale on Alex Jones's *Infowars* platform. Purveyors of these products have found accepting audiences[22] there. Such supplements and other "alternative" medical treatments have previously been the near-exclusive domain of left-wing thinking. Like the anti-vax movement, that marketplace has now turned purple, giving two things the radical red and the radical blue communities can now agree upon.

No matter what a politician says or promises during a campaign or even while in office, the most fundamental measure of political support is how much money from the federal budget gets allocated to the cause. It so happens that since the end of WWII, when science investments became an enhanced priority, funding to the White House Office of Science and Technology Policy—the branch of the US government overseen by the president's science advisor (now secretary of science)—combined with other non-military R&D, including agriculture and transportation, has increased slightly more under Republican than under Democratic presidents.[23] Worth noting: the highest budget gains went to Republican Eisenhower (46 percent increase per year during his two-term administration). Second place goes to the Democratic Kennedy-Johnson White House (39 percent increase per year; the Apollo years of the 1960s). During Trump's term, the budget increased 2.4 percent per year. The two lowest go to Democrats Clinton (2.2 percent increase per year) and Obama (1.2 percent increase per year) across each of their two-term administrations.

The political "science deniers," by this measure, actually do like science.

TROPE 3: Republicans are racist, sexist, anti-immigrant homophobes. Democrats embrace all peoples.

This label-laden trope is how Democrats see Republicans relative to themselves. In the past, the opposite prevailed.

Abraham Lincoln was the first Republican president—a party birthed, in part, to abolish slavery in America. Through Reconstruction and beyond, Republicans led congressional movements to finance and support the rise of Historically

Black Colleges and Universities, especially via the second Morrill Act of 1890, at a time when private elite colleges denied admission to all people of color. A summary web page on the history of HBCUs by the Smithsonian Institution[24] makes no mention of Republicans in securing these opportunities during post–Civil War America. Maybe they are trying to be non-partisan. By doing so, they are hiding the blunt and extraordinary fact that for 100 years the most racist political party was the Democrats. They oversaw the Jim Crow South and turned a blind eye to the thousands of lynchings[25] that occurred there. The governors, the mayors, the police chiefs, the angry chanting mobs across the region, as seen in the ugliest footage of the Civil Rights movement, were all Democrats.

Today, these tropes have all but switched, with a 180-degree realignment of who's inclusive and who's not. Except it's not entirely 180 degrees. Since 1990, the first two Black secretaries of state, Colin Powell and Condoleezza Rice, were appointed by Republican presidents. The second-ever Black Supreme Court justice, Clarence Thomas, was also appointed by a Republican president. Yet none of the four appointed by Democratic presidents Clinton and Obama were Black.[26] If you don't happen to like their politics, then confess you instead sought a high-ranking Black person who aligns with your political party's views, and not simply a high-ranking Black person. Add to this the landmark 2014 Pennsylvania case *Whitewood v. Wolf*, which declared the state's ban on same-sex marriage to be unconstitutional. Who presided over the case? Our old friend, Bush-appointed John E. Jones III. Continuing this round-robin, in April 2022, Judge Ketanji Brown Jackson became the first Black woman on the US Supreme Court, with fifty out of fifty Democratic senators voting for her and forty-seven out of fifty Republican senators voting against her. All this leaves me wondering what it means to be aligned with a political party at all. Do they do your thinking for you? Do they define your attitudes toward issues that confront the country? If so, then you are a pawn of those in power. A sentiment that resonates with a favorite lyric from "Sir Joseph Porter's Song," the admiral of the Queen's Navy, in Gilbert and Sullivan's comedic light opera *H.M.S. Pinafore* (1878):

I always voted at my party's call,
And I never thought of thinking for
myself at all.

But in a representative republic, those in power should be a pawn of you: government of the people, by the people, and for the people.

Views of Earth from space transform global perspectives for the better, I would say. But evaluating and judging individual humans from a distance hardly ever ends well. The brushstrokes with which we paint and characterize the views of others tend to be broad and without nuance, leaving us susceptible to bigotry and prejudice. From afar, a suburban lawn is simply a green carpet. When viewed close up, the carpet resolves into individual blades of grass. Closer still, the blades of grass resolve into plant cells that undergo photosynthesis. At what distance will you choose to formulate your opinions and perspectives about the lawn beneath our feet?

The 1980 season of *Cosmos*, hosted by Carl Sagan, was presented by the Los Angeles PBS station KCET. *Cosmos* is a multi-part science documentary, so naturally it landed on PBS. For the 2014 season of *Cosmos*, I was privileged to serve as host. This time around it premiered on the Fox Network, which happened to grant us the freedom and resources to create the stories and topics that compelled us.

My most left-leaning friends tend to view all of Fox as a Fox News monolith. Upon learning that *Cosmos* would not only not appear on PBS but would instead premiere on Fox, they presumed that Fox would dictate conservative party agendas to us, forcing us to be a mouthpiece of divisive Fox News ideologies. The less liberal of my friends had fewer of these thoughts, whereas those in the middle of the political spectrum congratulated us for securing a broadcast platform for science that was vastly larger than the PBS viewership.

Why this range of reactions?

Those farthest left were blinded by their own bias, tainting their capacity to see the world rationally. The politics of Fox News commentators incenses them. It incenses me as well. But in their far-left worldview, everything Fox was synonymous with Fox News. They never noticed that entire swaths of Fox's portfolio are paragons of progressive programming. Highlighting just a few examples, Fox is 20th Century Fox, which brought *Avatar* (2009) to the screen—a sci-fi blockbuster (the single highest grossing film of all time) that chronicles the plight of indigenous peoples on another star system, where they harness the mystical powers of plants and woodland creatures to defend their native planet against greedy corporate colonists. It might well have been titled *Pocahontas in Space*.

Searchlight Pictures is the indie production studios of Fox, which brought *Slumdog Millionaire* (2008), *12 Years a Slave* (2013), and the Oscar-winning documentary *Summer of Soul* (2021) to the screen, each an exploration of the plight of the disenfranchised. Fox is Fox Sports, which is highly regarded worldwide for their expert, thorough, tech-savvy, and diverse coverage. Fox is the cable channel Fox Business, which carries some Fox News DNA, but they are mild by comparison.

Most importantly, Fox is also the Fox flagship channel. Home of the acerbically liberal *Simpsons* and *Family Guy* as well as my personal favorite *In Living Color,* which was sketch comedy with a social conscience. These shows, and many more, broke new ground for their progressive social commentary. For example, *Glee*, a musical comedy-drama that ran for six seasons, featured the social exploits of a high school glee club. In one scene, two cast members sing a holiday favorite intended for a man and a woman. But this duet was sung by two men in love—with each other.

You can see how disappointed I was to suffer through righteous liberal lament over the presumed demise of *Cosmos* for appearing on the Fox network.

Hidden bias can cause a persistent urge to see all that agrees with you and ignore all that does not, even when countervailing examples abound. Among the many categories of how to fool oneself, the most pernicious is confirmation bias: you remember the hits and forget the misses. It affects us all, at one level or another. The antidote? Dispassionate rational analysis.

TROPE 4: Republicans are true patriots. Liberals are anti-American, and all they want to do is raise taxes and live off government social programs.

Back in 1781, the state of Massachusetts was the first to recognize July 4, Independence Day, as a holiday. Only six years earlier, Massachusetts had hosted the first battles of the Revolutionary War that created the United States of

America. That was long ago, but worth a shout-out for this bluest of blue states.[27]

Liberals and progressives planned and led nearly all the post–WWII anti-war marches. Is it anti-American to be anti-war? Liberals do like to ban things, almost always on the grounds that what needs to be banned is bad for you or for the environment. So maybe they're not trying to restrict your freedoms; instead, they're just trying to save your life.

What about taxes? Before you align yourself politically on this platform, consider actual reality informed by data instead of an imagined reality fed by incessant repetition. Rank the fifty states by federal tax revenue per capita paid in any given year. This can correlate with the state's economic health, but that alone is not what matters here. Next, include the total federal outlays per capita received by that state. The difference between these two quantities directly measures how much a state depends on government programs to function, and how much the government depends on the state to run.

When you conduct this exercise, you find that eight of the top ten states that pay more per capita to the federal government than they receive are blue states. On the other end, not including Virginia (home of the high-budget Pentagon), six of the ten states that receive more support from the federal government[28] than they pay are red states.[29] Given prevailing political rhetoric, you might have expected zero out of ten. Yet, under liberal Democrat presidencies, taxes levied have increased more than under conservative Republican presidencies. The "tax-and-spend" accusation is real: if you don't want to pay extra taxes then don't vote Democrat, even though anti-tax red states benefit greatly from increased tax revenue. The health and wealth of the nation remains highly dependent on the economic strength of blue states, with New York, New Jersey, Connecticut, and Illinois leading the way.

~~

Is there a world with no Democrats or Republicans? A world with no left-wing or right-wing zealots? Can we create a world of peace, with no war and no bloodshed, dotted only with mildly arguing people, who still want to have a beer with each other when they're done disagreeing over subjects that have no foundation in objective truth? On a landscape of endless political strife, if a peace-loving space alien landed on Earth and walked up to you and requested, "Take me to your leader!" would you escort it to the White House or to the National Academy of Sciences?

At the risk of overanalyzing fantasy, allow me to report on the culture of ComicCon. In San Diego, California, and in New York City, scads of people descend annually on those cities' convention centers to celebrate the world of cosplay, comics, animation, fantasy storytelling, superheroes, computer gaming, aliens, and especially science fiction on television and in film. They like to build and live in artificial worlds with self-consistent rules. They then like thinking rationally within those worlds. Combined, these two independently organized ComicCons, the largest two on Earth, draw more than 300,000 people.[30] Around the world, the total attendance at similar conventions may reach millions.[31] Attendees are diverse in every way. Tall, short, thin, chubby, disabled, gender-ambiguous, on the autism spectrum, bespectacled, unkempt. Many have never won a popularity contest and would never be a contender for homecoming king or queen, though they probably got higher grades than other kids in their school. I suspect (but cannot prove) that the Venn diagram of ComicCon'ers completely contains the circle of every person who has ever been wedgied by high school bullies in the history of the universe.

Everyone comes together out of a common love of imagination—arguably something that runs deep in our collective DNA. Yet there are no judgments.

No, that's not true. Of course there are judgments.

For example, the penalty is harsh but temporary if your R2D2 costume from *Star Wars* is missing its octagon port. Are your sword and pleated leather skirt believable in your Xena Warrior Princess outfit? Is your imitation of a foot-dragging zombie convincing? Does your handheld *Star Trek* phaser make the noise it should? If not, people will talk geek-smack about you. Beyond that, there are no judgments. Based on all I know of the community, being a card-carrying geek myself, and having attended many ComicCons on both coasts, I can confidently assert that attendees are largely scientifically literate. They long for all the ways the future of science and technology can transform the world (the universe) into a better place. They can distinguish between fantasy and reality—most of the time. They always know the difference between good and evil, and crucially, they live and let live. If ComicCon'ers ran the world, the worst geopolitical fights would be fake light-saber battles after a Friday lunch at the United Nations commissary.

Instead of the White House, why not take our visiting space alien to ComicCon. We'd have legitimate concerns that nobody would notice an actual alien camouflaged among those pretending to be one. The upside? Our alien visitor phones home and instead reports—"They're just like us!"

FIVE

RISK & REWARD

Calculations we make daily with our own lives and the lives of others

To understand probability and statistics is to understand risk—something the human brain is not wired to embrace intuitively. Consider that arithmetic, algebra, geometry, trigonometry, the graphing of formulae, logarithms, imaginary numbers, number theory, and calculus were off and running before anybody demonstrated it might be a good idea to take an average.[1] Arab mathematicians from the Golden Age of Islam, especially Ibn Adlan (AD 1187–1268) more than seven hundred years ago, began thinking about sample size and frequency analysis, establishing probability theory's earliest concepts, but a full-bore treatment of the field would not arrive until the 1800s.

The nineteenth-century German mathematician Carl Friedrich Gauss is considered by some (myself included) to be the greatest mathematician since antiquity. Shortly after the first-ever asteroid, Ceres, was discovered in 1801, its orbital path was tracked with spotty observations before it was lost in the glare of the Sun. How do we find it again when it emerges on the other side? Gauss decided to help out and develop the statistical method of "least squares"—the best mathematical way to fit a line through data, allowing you to predict what the data may do next. That tool allowed Gauss to predict the spot in the sky where Ceres would show up. As indeed it did. Right on time. Right on place.

By 1809 Gauss had fully derived the famous "bell curve," perhaps the most powerful and profound statistical tool in all of science. Also known as the "normal distribution," it reveals that for practically everything you might measure in the world, most values you report will fall in the middle of a range. At either higher or lower values, fewer and fewer examples of those values appear. This feature is especially true for the uncertainties that arise from measurements themselves, but also for quantities that may have actual variation. For example, few people are very short. Few people are very tall. Most people land somewhere in between. The concept is not more complicated than that, but the precise mathematical expression of the bell curve has made strong men weep:

$$f(x) = \frac{1}{\sigma\sqrt{2\pi}} e^{-\frac{1}{2}\left(\frac{x-\mu}{\sigma}\right)^2}$$

Yes, it sports three lowercase Greek letters—count them: *sigma*, *pi*, and *mu*. It also has a fancy italic *f* and the exponential function *e*, all in one equation with *x* as the variable. When plotted, the curve takes on the shape of a bell. Not a sleigh bell. Not a cowbell. More like the Liberty Bell.

Well before this equation appeared, foundational physics for landing on the Moon had already been established, and the industrial revolution was in full swing. Further evidence that thinking statistically about the world is not only unnatural, advances in the field required some of the smartest people who ever lived. We also confront a curious modern fact that many leading universities have a Department of Statistics separate and distinct from their Department of Mathematics. Yet you do not find separate departments for other branches of mathematics. No Department of Trigonometry. No Department of Calculus. Evidence that statistics is different, and somehow requires its own thinking space.

When statistically unlikely events occur—at random—adults commonly draw from a huge reservoir of meanings to account for them. The need to do so, coupled with an overall absence of curiosity for what is true, may have rational evolutionary roots.[2] For example, is that a lion rustling the tall grass ahead of you, or just the effects of a gentle wind? Consider the outcomes in a hungry-lion flowchart:

1. You think you see a lion. You are curious and you want to make sure, so you walk closer and discover it's indeed a lion. The lion then eats you, summarily removing you from the gene pool.

2. You think you see a lion. You are curious and you want to make sure, so you walk closer, only to discover it's just the breeze. You live another day. But keep up this behavior and you eventually suffer outcome No. 1.

3. You think you see a lion. It's actually a lion. You ran away before you could confirm this. You live another day.

4. You think you see a lion. It's not a lion. It was just a breeze. You ran away before you could confirm this. You live another day.

Notice who's rewarded here: those who saw patterns, whether or not they were real, and those who had no curiosity.

Our ancestors were also highly dependent on assumptions of cause and effect to navigate survival. If you're eating some berries in one moment and get very sick in the hours that follow, the cause was probably the berries. The coincidence of these two events weighed heavily on our understanding of the world. Those who didn't make the connection kept getting sick and faded from the gene pool.

Even though no lions lurk behind parked cars and no poisonous berries await us at the corner grocer, these prehistoric behaviors, when ported to modern civilization, remain with us and manifest across a broad spectrum of irrational behavior.

For example, in a chance encounter with a long-lost friend in a far-off place, often we think it was preordained, possibly declaring, "There are no coincidences!" If not, we might otherwise make the geographically questionable claim, "Small world!" But try approaching every person you see in the street and asking, "Do I know you?" When they answer "No!" declare out loud "Large world!" Spend a single day doing this and you'll never proclaim "Small world!" again. In another example, how many of us don lucky socks or underwear on important days? They became lucky because that's what you happened to be wearing when something unexpectedly good occurred in your life.

In another example, brought to you by advertisers, they know in advance that offering you the statistics of their product's effectiveness will be ineffective. So they infuse their commercials with compelling testimonies of people who look just like you, and who declare how splendidly the product served their needs. We are more likely to be swayed by a single person who testifies with passion than by a bar chart containing data compiled from thousands of people.

The urges to think in these ways are strong and normally harmless. But our shortcomings are well known and shrewdly exploited—hijacked—for economic gain by casinos and other centers of gambling. Imagine how different the world would be if thinking mathematically about human affairs was normal and natural. Such powers of analysis would influence nearly every decision we make in a day—especially decisions that may influence our uncertain futures. There is no analysis of scientific data, especially in the physical sciences, without tandem and thorough multi-year undergraduate and graduate-level courses in probability theory and statistics to support it. For these reasons above all, the world looks very different to scientists.

Scientists are human too, but the extensive mathematical training slowly rewires these irrational parts of the brain, leaving us a bit less susceptible to exploitation. In one shining example, consider the American Physical Society (APS), which is the primary professional organization of the nation's physicists. In 1986, due to hotel scheduling conflicts, they were forced to cancel plans at the last minute for San Diego to host their annual spring meeting. On just a few months' notice Las Vegas became a fast-and-easy substitute, and the MGM Grand Marina became the lucky host of 4,000 physicists.[3] This hotel, now in a new location with nearly 7,000 rooms, was and is the single largest hotel in the USA. With more than three acres of casino on the property,[4] their business model is not hidden.

Guess what happened.

That fateful week, the MGM hotel earned less money than in any previous week—ever. Could it be that physicists know probability so well that they boosted their odds against the casino in poker, roulette, craps, and slot machines and came away victors? No. They simply didn't play.

The physicists were inoculated from gambling by mathematics.

PHYSICISTS IN TOWN, LOWEST CASINO TAKE EVER

1986 Las Vegas newspaper headline
Note: Subsequently, Las Vegas asked the APS to never return to their city.

Compared with other Earth-based applications of probability and statistics, casinos specifically and perniciously target our weaknesses. Just because your favorite number, say 27, has not come up in a while on the roulette wheel doesn't mean 27 is "due." Each spin has no memory of any previous spin, leaving you with the same odds on every spin. Yet every roulette table lists the results of the previous dozen or so spins, just to feed our ignorance of how probability works. Our primate brains simply can't handle this truth.

A few more of these. Opposite sides of a proper die will always sum to seven. Six and one. Five and two. Four and three. Seven is also the most likely roll on a pair of dice. Lucky seven. But rolling a seven is still unlikely. On average, five-in-six rolls will not get you seven. How about eleven? That's a 1 in 18 roll of the dice. Things to know before you willingly, if unwittingly, allow a casino to take your money.

If you happen to be on a rare streak of wins—winning intermittently is precisely what feeds the addiction—the casino takes note and sends over a comely server to offer you alcoholic drinks on the house. Just what you need in that moment, a means to further distort your capacity to think.

None of this subtracts from those who simply like to gamble every now and then. When in Vegas, I like betting various combinations of 2, 3, 5, 7, 11, 13, 17, 19, 23, 29, and 31 on a roulette table. That's the wheel's full supply of mathematical prime numbers. Statistically, they're as good (or bad) as any other set of eleven numbers you might pick. If I'm going to hand my money to a casino, I'm going down while doing some math. I typically allocate about $300 and make it last several hours. Upon returning from the casino, when people ask how much I lost, I reply that I gained $300 of entertainment—about the cost of dinner, wine, and opera in my hometown. Curious, then, that nobody asks, when you return from the theater, "How much did you lose?"

In the United States, organized gambling is ubiquitous. Casino revenue for 2021 hit an all-time high of $45 billion.[5] That's nearly twice NASA's annual budget to explore the universe. Forty-five out of fifty states offer some kind of a lottery,[6] including Powerball, in which the public spends nearly $100 billion annually hoping to win the jackpot—or at least win more than they've spent on their tickets. As you'd expect, the larger the jackpot, the more lottery tickets get sold. Buying more tickets does indeed increase your chances of winning, but jackpots are normally shared among winners, so statistically your winning take dilutes as the number of ticket buyers increases.

In a recent contest, the odds of winning the Powerball jackpot in Tennessee were 1 in 292.2 million.[7] Many people take those odds, hoping—even expecting—to win. Although you are twice as likely to be struck and killed by lightning. Yes, that means your tombstone will more likely say "Killed by Lightning" than "Won the Tennessee Powerball Lottery." States that have otherwise outlawed casinos have nonetheless blessed gambling when run by their own legislature.

While we're in Tennessee, imagine someone named Claire who wins their jackpot. She says she's good at predicting future events. Even though her last name is Voyant, here's a headline you are not likely to see:

CLAIRE VOYANT, TOWN FORTUNE-TELLER, WINS LOTTERY . . . AGAIN.

Her probability of winning it twice is 1 in 292.2 million times 1 in 292.2 million. That's 1 in 85 quadrillion. I'm just saying.

The best justification I've heard for playing the lottery was from the mother of an astrophysics colleague. She occasionally buys a single weekly ticket, and during those seven days, awaiting the draw of numbers, she browses those fancy real estate brochures featuring beautiful homes that hardly anyone can afford. She fantasizes about living in a home of her choosing, and these longings bring her temporary joy, worth the price of the ticket. Who am I to stop her?

Profits earned by the state, after paying winners and ticket merchants, serve as a major source of revenue that often funnels into social programs, especially kindergarten through high school education, creating a moral dilemma to vote against this form of legalized gambling in your state. That got me thinking. Is probability and statistics even taught in US public schools? Recent surveys[8] show that the answer is mostly not. In the few places that do teach it, classes are taught as a novel elective or as part of an advanced placement college course. If instead, probability and statistics were a fundamental part of the K-12 curriculum, taught to every student, across multiple grades, and if state lottery revenue were allocated to make that happen, then the lottery might just put itself out of business by inoculating its own citizens against the lottery itself.

Some years ago, while walking through Las Vegas's McCarran Airport, I did the vain-author thing and stopped by the bookstore to see if one of my recently published books was on display. I would then volunteer to sign their stock, increasing the chances they would make a sale.

I couldn't find the book, but had only glanced at the shelves and might have missed it. Plus, airport bookstores are tiny. And the book was not a bestseller, so there's no expectation for them to carry my book. I nonetheless tactfully asked the cashier, "Where is your science section?" The reply was simple and direct, "Sorry, we don't have a science section." My silent reaction in that moment became my inaugural Tweet[9] among thousands to follow that would capture my random daily thoughts as an educator and as a scientist—the world through the lens of an astrophysicist.

Neil deGrasse Tyson
@neiltyson

Borders Books at Vegas airport does not have a science section. Wouldn't want to promote critical thinking before you gamble.

3:46 PM · Feb 9, 2010 · Twitter Web Client

If visiting space aliens analyzed what's going on here, they might wonder what kind of a species would purposefully exploit the frailties of its own kind, creating a systematic transfer of wealth from the gambler to the casino owners, be they in Vegas or in the state capitol.

Good evidence for no sign of intelligent life on Earth.

Some of these irrationalities derive from the urge to feel special—a benign force that looks after you, making unlikely things happen in your favor. Another thought experiment: Line up 1,000 people and get them to all flip a coin. For an ordinary coin, with a 50 percent chance of landing heads or tails, approximately half of the thousand will get tails. Ask them to sit down and have the remaining 500 people continue the experiment. Tell the 250 people who got tails to sit down, just as the first 500 did. The numbers will vary slightly from one experiment to the next, but on average those left standing will halve their way from 1,000 to 500 to 250 to 125 to 62 to 31 to 16 to 8 to 4 to 2 to 1. This outcome is obvious, but let's look more closely. After five coin tosses, about 30 people will have flipped heads five consecutive times, eliminating 970 people. How about the last one standing? That person flipped heads ten consecutive times. This has never happened in your life, yet it will occur to some person most times you repeat this experiment. Who does the press rush to talk to? Not the 999 losers, but the 1-in-1,000 who got ten heads in a row. You can imagine the conversation:

Eager Reporter: Did you think you were going to win?

Happy Winner: Yup. This morning I felt some heads-energy in the room. Halfway through, that feeling increased. With a few flips left, I knew I was going to win.

During that brief exchange, our fictional flipper converted a completely random statistical outcome into mystical destiny. If you think this experiment is too unreal to be relevant, consider the stock market. At the end of a trading day (or week or month), you can expect only two real outcomes of any market index or investment vehicle you care about. Could be the Dow Jones Industrial Average, the NASDAQ composite, tech stocks, cryptocurrency, municipal bonds, pork bellies, it doesn't matter. The investment will end the day traded lower or higher than the day before. It could also remain unchanged, but that's irrelevantly rare in this example. One more bit of blunt reality: on the expectation that the price will drop, you sell securities to a person who buys them on the expectation the price will rise.

No matter what the market does in a day, the news give reasons for it. Even small fractional day-to-day changes get explanations to account for them. Occasionally they offer no reason at all, not even implicit befuddlement. Consider this highly typical headline of the investment universe that the financial TV network CNBC Tweeted on December 10, 2021:[10]

> THE MAJOR AVERAGES ROSE ON FRIDAY EXTENDING WALL STREET'S STRONG RALLY THIS WEEK, DESPITE INFLATION HITTING A 39-YEAR HIGH

If they were honest, here's what that headline would have said:

> THE MARKET ROSE TODAY. WE HAVE NO CLUE WHY AND REMAIN DUMBFOUNDED.

To probe this further, line up a thousand Wall Street analysts. There's many more than that,[11] but let's just stick to a thousand. Maybe some are better at making money than others. Let's not take that away from them. They might be good at predicting cultural trends and at navigating the countless simultaneous variables that can affect their portfolios. That almost always pays off. But let's pretend for a moment that the investment marketplace is completely random. If so, even if they used darts to guide their investment strategies, one out of a thousand analysts will correctly predict each day's outcome across ten consecutive days. Just as in our heads-or-tails experiment, a mere five days earlier around 30 market analysts would have correctly predicted the market outcome for five days in a row. That's only 30 left standing out of 1,000. If interviewed, the successful 30 and especially the last one standing will surely boast of special market insights. We will believe them, because the performance looks and feels impressive to both the investor and the analyst, all for something completely random.

Is the nation's most successful trader in one year also the most successful trader the next year, and the year after that? This hardly ever happens. On one of many sites that rank traders,[12] in response to my request for past data, they declared: "There is no way for you to see the historical rankings of experts on the site." So I took notes, waited a mere five months, and reviewed their list again. Of the top-ten ranked analysts in July 2021, none were still in the top ten. Investment firms know this shortcoming and legally alert you in the small print that "past performance is not an indication of future results."

If the last person standing happens to be the same each time, then something extraordinary is happening. We want such people to exist. We need such people to exist. They're evidence that the world is knowable and not random. That's good because we don't understand random. Warren Buffett's Berkshire Hathaway has done well over the past half century, although since 1965 it has ended eleven years in negative territory, two of them bad. In 1974 its value dropped nearly 50 percent and in 2008 by more than 30 percent.[13] What we really want is a consistent winner. One that does not foster market anxiety from year to year. Such a person did exist. His name was Bernie Madoff, with a decades-long winning streak that defied all odds. He must be good. Or he must be cheating. Or he must have been good at cheating. Madoff cloaked nearly $65 billion of people's savings in the largest Ponzi investment scheme ever exacted upon an unsuspecting public. Convicted in March 2009, he died while incarcerated in April 2021, long before fulfilling his 150-year prison sentence.

Some say the stock market is the world's largest casino. I largely agree, except nobody brings you free drinks.

Even when we're not visiting Las Vegas, probability figures in everyday decisions we make. Consider the public's sentiment toward genetically modified organisms—GMOs. Reactions tend to be bimodal, depending on your politics, itself a warning flag. The truth and efficacy of science should never correlate with your political views. Left-leaning people tend to see GMOs as an evil, unhealthy scourge on health and civilization. Scientists and right-leaning people[14] tend to be just fine with it. A full discussion of the topic falls far outside the scope of this book, although I did narrate a documentary[15] that explored the science of GMOs as well as the cultural and political divide it has caused. Here, I instead offer a statistical anecdote to whet your appetite.

The food chemical company Monsanto, now owned by Bayer, developed a genetically modified variant of corn that was completely resistant to glyphosate, a weed-killing herbicide marketed under the name Roundup, which they also developed. Monsanto scientists genetically removed their corn's susceptibility to the chemical. This potent combo—Monsanto's GMO corn coupled with Monsanto's weed killer—enabled farmers to spray their entire crops and have the herbicide kill everything but the corn. The Vermont ice cream company Ben & Jerry's uses corn syrup as a sweetener for some of their products. (Yes, I too was surprised to learn this.) News that some of their ice creams had trace amounts of glyphosate from the corn used in their syrup created a media dust-up. In response, Ben & Jerry's decided to stop using GMO corn syrup altogether,[16] even though the one-part-per-billion detection levels of glyphosate were far below US and European standards. Since many people who buy Ben & Jerry's ice cream lean left—aligned with the company's generally progressive views on all things—Ben & Jerry's Homemade Holdings Inc. judged this ban to be a wise business decision.

Let's look closer at what happened there. Every substance you could possibly ingest, food and otherwise, has a calculated lethal dose associated with it, measured by what's called LD50. That's the dose per kilogram of body weight where 50 percent of the people who consume that amount will die quickly. These data often come from tests on laboratory mammals such as mice. There's another metric, called no-observed-adverse-effect level (NOAEL), which addresses the long-term influence of a substance on your health and is more sensible when thinking about food safety. LD50 helps to make a different point. The smaller its value for a substance, the more lethal it is. As such, tables of LD50s can be quite illuminating. Here's a sampling:

Sucrose (table sugar)	30 grams per kilogram
Ethanol (common alcohol)	7 grams per kilogram
Glyphosate (Roundup)	5 grams per kilogram
Table Salt	3 grams per kilogram
Caffeine	0.2 grams per kilogram
Nicotine	0.0065 grams per kilogram

The most lethal substance on this hand-picked list is nicotine. Caffeine looks quite potent too. Just drink about a hundred and fifty demitasse cups of espresso if you want to die from it. Next comes salt. Clearly, then, being complimented as the "Salt of the Earth" cannot always be a good thing. This famous line from *The Rime of the Ancient Mariner* implicitly captures salt's LD50, contemplated by the thirsty mariner, surrounded by salty ocean water: "Water, water everywhere, Nor any drop to drink."[17]

The least deadly on the list is sugar, as you might expect. Notice further that glyphosate is less lethal than table salt, but not by much. Actually none of this concerns us here. What matters is what happens to a 150 lb. (70 kg) person who eats Ben & Jerry's ice cream—a fact I calculated but relegated to my Forbidden Twitter file, where it remains, simply for how disturbing it would be. In social media, I never intend to be disturbing:

> *You would need to consume four hundred million pints of Ben & Jerry's ice cream for its trace amounts of glyphosate to kill you. But after only 20 pints you will die from its sugar content.*

Ben & Jerry's made the right corporate decision if it protected their profits. Although they could have also used the occasion as a teaching moment—a mind-blowing lesson on comparative risk. But that works only if people are open to learning. In modern times, many of us don't satisfy that criterion, perhaps because, according to the nineteenth-century British essayist Walter Bagehot,[18]

> *One of the greatest pains to human nature is the pain of a new idea.*

But more of his quote tells it all:

> *It is, as common people say, so "upsetting;" it makes you think that, after all, your favourite notions may be wrong, your firmest beliefs ill-founded. . . . Naturally, therefore, common men hate a new idea, and are disposed more or less to ill-treat the original man who brings it.*

Another overlooked dimension of risk is our willingness to embrace studies that tell us our habits or diet may increase our chances of contracting cancer. Often, when such studies are reported, they tell you how much your risk of cancer increases when you engage in one kind of activity or another. Knowing the baseline risk for that particular cancer is paramount, yet we hardly ever pay attention to that statistic. For example, let's analyze this sentence from the American Cancer Society's web page[19] on colon cancer: "Cooking meats at very high temperatures (frying, broiling, or grilling) creates chemicals that might raise your cancer risk." The word "might" appears because some studies show no increased risk at all. In any case, I happen to like searing meats at very high temperatures, but I also don't want to get cancer. The web page offers an entire discussion of multiple risk factors, but does not quantify my baseline risk nor say by how much that risk is increased. Digging elsewhere, however, I learn that my lifetime risk of getting colorectal cancer is 4.3 percent.[20] And from a separate meta-study of research articles[21] I learn that my increased risk of colorectal cancer on that baseline is about 15 percent, with huge variance from study to study. Nobody wants their chance of getting colorectal cancer to increase at all, let alone by 15 percent. What's mathematically clear, but conversationally deceptive (especially if you just read headlines) is that your lifetime chance of contracting colorectal cancer has not increased by 15 percent. What has increased by 15 percent is the risk on your baseline risk. If you eat meats grilled at high temperatures, then your lifetime risk increases by only 0.6, from 4.3 percent to 4.9 percent, which is indeed a 15 percent increase.

If you're a barbecuing meatarian, you can choose to accept or reject this increased cancer risk in your life. We simply need honesty and transparency when reporting these statistics if we are to make informed lifestyle decisions.

Another challenge for the human brain to grasp is slow existential threats. They're easy to deny, often because the danger is neither clear nor present. If you smoke heavily, for example, you're surely aware that you face increased risk of death from lung cancer or related heart diseases. But it's your body. It's your cigarette. Dammit, it's a free country. So you accept that there's a 1-in-8 likelihood[22] your tombstone will read "Died from smoking."

Just to be clear, you're betting against an outcome that

carries more favorable odds than most casino bets on outcomes you seek.

With the help of yet another thought experiment, let's speed things up a bit. Same risk as before except we accelerate the timeline and add some gratuitous gore. All regional authorities designate next Tuesday as "Cigarette Smokers' Day." The first puff taken by one out of every eight smokers, at random, will cause their skulls to explode, leaving them as collapsed, headless, bloody corpses on the pavement. If you happened to remain alive that day, you could smoke for the rest of your life and die from some other cause.

On that fateful Tuesday, the streets and smoking lounges of America would be strewn with four million headless bodies—three times the death toll the US sustained in all its armed conflicts combined, including both world wars, Korea, Vietnam, and the Civil War. A gory day indeed, but the exploded-heads scenario would be far less costly to society since that mode of death incurs no protracted medical bills from trying to keep terminal cancer patients alive.

If you love to smoke, would you take that risk?

When you explore the same basic information—the same data—from many different perspectives, especially when you compare one risk that you accept to another that you reject, the relevant details shine brightly while the irrelevant details melt away. These are the beginnings of an enlightened, scientifically literate perspective.

~

How about safety? We all want to live long and prosper. What about the overall risk of dying prematurely from all causes if you live in the city versus the suburbs? Big cities have always been a hotbed for crime and homicide, but that's where the businesses are. So why not live in the city for now, get married, make a bunch of money, then move to the safety of the suburbs to raise a family. That's what the suburbs are for: a means of escaping everything bad about city living.

A prime example of wishful, selective thinking.

If that's your reasoning, your fantasy has overridden your search for conflicting data. Holding aside that nearly all mass shootings in schools happen in the suburbs,[23] if you add up the lethal risks to life in the city versus life elsewhere, turns out you're safer in the city.[24] The causes of possible harm are different, but enlightening to compare. In the suburbs, traffic fatalities are much higher than in the city, as are overall accidents (drowning included), suicide, and drug overdoses. All combined, on average, your chance of dying prematurely in the suburbs is 22 percent higher[25] than in the big city.

This analysis simply required that you step back from assumed truths, gain a wider perspective, and query the data in different ways, none of which is possible in the tunnel vision of bias.

Regarding mass shootings, I once posted a Tweet that should have been relegated to my Forbidden Twitter File, but I mistakenly thought people would be comforted to know that mass shootings are a tiny fraction of all preventable deaths in the country. Mass shootings are even a tiny fraction of all gun deaths, and emotions more than data drive our reactions to them. My Tweet posted within days of the 2019 El Paso, Texas, shooting,[26] in which 46 people were shot in a Walmart, 23 of them killed. I was instantly pilloried in social media for my insensitivity to the victims and their loved ones.

Years earlier, but well after the events, I had made a similar point about the four-plane American death toll from the terrorist attacks of September 11, 2001. Nearly 3,000 people died that day, all of whom expected to come home for dinner. I noted that we lose about 100 people a day to traffic fatalities, which means that by October 11, 2001, one month later, we had lost more people than died on September 11. That statistic continues to accumulate, month by month, and will not abate until we do something about it. Every year we continue to lose upwards of 35,000 people on our roads, yet the US military has spent $2 trillion on our post–9/11 war on terror,[27] mostly in Iraq, precipitated by the singular deaths of September 11. America was angry and did not want to live in a state of terror. This wasn't a cost-benefit calculation about saving lives. It was a cost-benefit calculation about how we feel.

In another example of facts versus feelings, consider proposed solutions to the meteoric rise in the deer population that wander in residential areas of the northeast US. Deer account for unending car accidents that result in human injury and death, not to mention astronomical insurance costs. One proposal to combat this hazard reintroduces the native species of large, deer-munching feline carnivores that once roamed the region.

What could go wrong?

A study from 2016, authored by nine wildlife scientists, modeled the predator-prey relation between cougars and white-tailed deer.[28] They reported that within thirty years, a vibrant predator population dining upon unwanted deer can avoid 21,400 injuries, prevent 155 human deaths, and save $2.1 billion, all from car accidents that won't happen. Naturally, cougars also occasionally eat people, especially small wayward children—the model predicts about thirty of them. So we have two choices: 1) Introduce hungry cats that eat thirty people in thirty years, or 2) don't introduce hungry cats, and leave car-deer accidents unabated that injure thousands, kill hundreds, and cost billions.

If the societal priority is to save lives, but the interpersonal priority is to value our emotions, then how do we balance these factors in our daily lives? Laws and legislation and national directives pivot on this. Dying in a car-deer accident, even in large numbers, can be thought of as nobody's fault. Yet getting eaten by a large cat put there by the government is abominable. Do we admit (confess?) to ourselves that we are not cold mathematical creatures, and then celebrate our feelings, knowing they wield the power to override our rational thought? Or do we suppress all that might confound a rational decision? Could or should we allow emotions to influence legislation in response to data?

As self-driving cars and other futuristic Jetson-like technologies gain presence in our world, we will encounter a similar dilemma. Human error causes more than 97 percent of all traffic crashes around the world.[29] Meanwhile, self-driving cars are never inebriated. They are never sleepy nor susceptible to road rage. Their reflexes are nearly instantaneous. They can see unlit obstacles at night. They can see through fog. They're never texting while driving, and even if they were, it wouldn't matter. Furthermore, on a road with just self-driving cars, if any one vehicle wants to change lanes—a source of many fender benders—your car simply shares this information with surrounding cars, and they politely allow this to happen. During this inevitable transition away from human-controlled cars, unforeseen software and hardware errors will surely lead to traffic fatalities. Each cause will likely happen only once, as engineers update software to prevent the same situation from happening again. This will systematically drop the self-driving fatality rate to near zero per year.

Self-driving cars may ultimately save 36,000 lives per year in the US. What do you do emotionally, legally, societally, if self-driving cars still manage to kill, say, 1,000 people per year? No journalist will profile and celebrate each of the 35,000 random men, women, and children who didn't die that year from car accidents. Even if they did manage to write such a piece, there's no solace for the loved ones of those who died. That's what breeds *New York Times* headlines such as:[30]

TESLA SAYS AUTOPILOT MAKES ITS CARS SAFER.
CRASH VICTIMS SAY IT KILLS.

Both parts of the headline are true, but we lack the capacity to embrace them simultaneously.

The US aviation industry has experienced precisely that trend over the decades. For example, in the 1990s, more than 1,000[31] people died in airplane crashes. The decade that followed, not including the terrorist crashes on September 11, 2001, saw half that many deaths. During the ten-year span from 2010 to 2019 (excluding charter, cargo, and private flights), eight billion passengers flew on commercial airplanes without a single crash,[32] although two died from other causes.[33] The National Transportation Safety Board studies each incident, fatal or otherwise, and commonly offers findings that improve the safety regulations for air travel. Even more impressive is that over the decades, airline travel has been growing. By the end of 2019 (pre-COVID-19), passenger travel on domestic airlines had risen 35 percent from 2000.[34] Had the fatal accident rate per takeoff and landing remained the same, the total deaths would have gone up each year as passenger travel increased. Since people tend to react to pure numbers rather than to pure statistics, the aviation industry would have been viewed by many as becoming less and less safe, even if the opposite was true.

In Jonathan Swift's 1726 classic adventure novel *Gulliver's Travels*, one of Gulliver's excursions takes him to a fictional island off the southern coast of Australia, populated by a race of intelligent, exquisitely rational horses called the Houyhnhnms—yes, that's spelled correctly. In the surrounding woods roams a hairy, smelly, irrational, species of human-ape called the Yahoos. Gulliver realized in conversation with these horses that to them, he was in every way much more like the Yahoos than like themselves.

As a geeky kid, I remember on first reading this story how I longed to be like the rational horses. Their thoughts: crisp and clear. Their decisions: reasoned and rational. When I got older, I discovered for myself that emotions are what drive feelings. The Houyhnhnms are cold and emotionless. Yet feelings are a feature, not a shortcoming, of what it is to be human. So feelings can and perhaps should affect our personal equations of risk versus reward even if doing so may leave us occasionally confused about whether we made the right decision. Something Joni Mitchell knew well in 1967:[35]

> *I've looked at life from both sides now*
> *From win and lose and still somehow*
> *It's life's illusions I recall*
> *I really don't know life at all.*

All I ask is to see accurate and authentic data, analyzed from all directions—free of bias and tunnel vision—before I layer my emotions upon it. In the end, we must live with the consequences of our decisions. After all input of facts and statistical analysis, our emotions may defy reconciliation with data. That's okay too.

SIX

MEATARIANS & VEGETARIANS

We are not entirely what we eat

In Western culture, meat eaters don't tend to have reasons or philosophies behind their food choices. They simply enjoy the taste of dead animals—breaded, fried, grilled, cured, barbecued, roasted, sous vide, and smoked. For some, eating meat is all they've ever known, and they can't imagine life any other way. Vegetarians, on the other hand, especially those who have converted, offer all manner of reasons for their food preferences. Most common among them are to improve health and to protect the environment. For others, it's the abject abomination of raising, killing, and eating sentient beings. Or as a minimum, the need to avoid eating life-forms that can experience pain. Even worms will writhe in response to being unpleasantly poked.

While most are silent on the matter, there's always that evangelical vegetarian who tries to lure away meat eaters.

Their carnivorous counterparts are rare, but there's no denying the cultural stereotype of the virile, meat-eating man. What comes to mind is an ad campaign by the Beef Industry Council featuring the actor James Garner in cowboy boots, accompanied by a deep-voiced tagline: "Beef. Real Food for Real People." In one of several television commercials, he flat-out rejects the veggies on his shish kebob, complaining they always fall off on the grill, while the meat stays firmly on the skewer. Next time, just skip the vegetables. James Garner would later suffer a stroke and ultimately die of coronary heart disease—at age eighty-six. If James Garner can't lasso you back into the carnivorous corral, then maybe Jesus can. Want to rebut multiple claims that Jesus must have been a vegetarian? Just read the book *What Would Jesus Really Eat: The Biblical Case for Eating Meat*, reviewed in, of course, *Beef* magazine.[1]

The largest species of animal ever on Earth is alive today: the blue whale, a mammal carnivore dining principally on a centimeter-size crustacean called krill—tons of it per day. The largest land animals today are also mammals and include the elephant, hippopotamus, rhinoceros, giraffe, water buffalo, and bison. They're all herbivores. Polar bears are on the list too, but of course they're carnivores. Grizzly bears are opportunistic omnivores and will eat anything they want, including humans.

The world's animals are a mix of carnivores, omnivores, and herbivores, with the words meatarian and vegetarian swapped in when referencing human animals. That's because carnivores eat only dead (or living) animals and herbivores eat only living (or dead) plants. Meanwhile, human meatarians commonly eat things other than meat, like dairy. As do vegetarians. At about 40 percent, India has far and away the largest percent and total population of vegetarians in the world,[2] due primarily to Hindu religious traditions that include the sanctity of cows. The UK is around 10 percent vegetarian. The US, 5 percent. Based on the rapid growth of plant-based meat substitutes in the US and the rise of authentic vegetarian options on restaurant menus, you'd think the US number would be higher. That 5 percent figure has remained low and stable for more than a decade. Even Argentina is 12 percent vegetarian, and they're otherwise famous for eating all steak all the time.

If you're vegetarian, but remove all cheese and eggs from your diet, as well as milk and honey, you're vegan. In the US, they're around 3 percent of the population,[3] up significantly from less than 1 percent a few decades ago, but still hovering in the low single digits. Add that to the vegetarian number, and you've got 8 percent of the US population who do not eat meat.

Most people of the Earth, like grizzly bears, eat whatever is served for dinner. Over the past fifty years, the world population has doubled, yet meat consumption has tripled,[4] which tracks closely the rise in wealth among nations that previously had no access to this expensive protein. Despite widespread pleas from vegetarians, Earthlings are eating more meat than ever before.

Perhaps the most storied carnivore is the wolf. The "big bad wolf" makes an appearance in the fairy tales *The Three Little Pigs* and *Little Red Riding Hood,* as well as in the Russian tale *Peter and the Wolf.* When the carnivorous wolf wants to eat the pigs or Little Red Riding Hood or Peter, or in the real world when wolf packs bring down a majestic elk, they're not being bad wolves. They are just being wolves. They kill with no concern for the pain and suffering of their prey. Same is true for fish—all fish. The answer to the question

"What do fish eat in the ocean?" includes "Other fish." Apart from the smallest of fish who eat plankton, not a one of them is an herbivore. That's what accounts for the persistent and systematic concentration of heavy elements like mercury and other toxic industrial pollutants in large marlin and swordfish, who sit at the top of the fish food chain.

When watching nature documentaries, I'm surely not alone as I cheer for the defenseless plant-eating animals being stalked by the sharp-toothed carnivores. It's not easy being green. We delight as a bounding impala darts sharply to the side while the less nimble cheetah tumbles by at 75 mph in a failed attempt to secure dinner. Yet, cheetahs gotta eat too.

Despite the natural state of predator-prey relationships among animals on Earth, the argument remains that animals are sentient and that we, as rational humans, have the smarts and the resources to avoid eating them, thereby respecting their neurological endowments over other life-forms. That's a potent pretense, even if all the food-slaughtered animals were happy their entire lives.

~

No matter what you eat, if you source foods locally, you minimize the transportation footprint, which may be better for the environment than a simple vegetarian diet that pays no regard to where the plants were harvested. Although that outcome depends on many factors whose efficiencies are continually in flux: Does the food move by boat, train, truck, or plane? How much food has spoiled along the way? Are the truck engines electric or internal combustion? How does the local electric company generate its power? And how arable does your region of the world happen to be?

Apart from these issues, the production of meat in the US is staggeringly efficient. Across all fifty states, for example, we consume nine billion chickens per year—a rate that's 3 times higher than the world average. If you do the math, you get a million per hour, with each chicken living anywhere from six to twelve weeks before slaughter. Yes, for every hour of every day the US hatches, raises, kills, distributes, and eats a million chickens. At a few dollars a pound in some retail outlets, chicken constitutes some of the most inexpensive protein you can consume in the marketplace. We're quite efficient at making cattle too, although they take more time—one to two years—before being brought to slaughter.[5] They also occupy much more space than chickens, not only physically, but on the ranch. Depending on the terrain, one grass-fed bovine will require many acres of land[6] to graze. Don't want them to graze? Instead, cram them into feedlots where they create mountains of manure and rivers of urine. The largest among these in the US packs 150,000 cattle into 800 acres.[7] When ready for slaughter, a single 1,200-pound animal will supply nearly 500 pounds of meat.[8]

Cattle are fully domesticated. There are no wild herds of Holsteins roaming the countryside. There are no feral gangs of Wagyu steer lurking in the hills. Modern cattle were genetically invented by humans via selective breeding of the now-extinct, oxlike Eurasian aurochs. The goal? To exquisitely design a biological machine that turns grass into steak—or into milk, if you prefer.

I posted a version of that sentence on Twitter, and some people nearly lost their minds. Most notably the American musician and animal rights activist Moby. In an Instagram post he scolded:

> *When one of your heroes breaks your heart. Neil deGrasse Tyson, really? You can Tweet that and make light of the unspeakable suffering experienced by the hundreds of billions of animals killed each year by humans? . . . For a smart physicist, Neil deGrasse Tyson, you sound like an ignorant sociopath.*

Moby's full post and my even fuller reply are elsewhere.[9] All that matters here is that my statement was a simple expression of fact, offering no opinion at all, with the added imagery of a "biological machine." Some people thought my Tweet was pro–animal slaughter while others thought it was a blunt appeal to turn everyone into a vegetarian. More evidence we carry filters that bias how we process neutral information. Moby has since apologized for the tone, but what's unforgettable here is the intensity of his well-founded activism.

The production of animal-based foods is the pride of assembly-line manufacturing in the industrialized world. Do you live elsewhere? The tree of life is well sampled for dinner, and many meals include horse, ostrich, emu, kangaroo, and dog, as well as reptiles and insects. Let's not forget rodents. In Texas I once ate barbecued squirrel. Still had some lead pellets in it from being shot. Had to pull those from my mouth one by one. After all that, not much meat in there. Of course, it tasted like chicken.

Fish don't moan or scream. They also don't drop quarts of blood when you cut them open. Perhaps that's why you hear less about their plight en route to your dinner plate. The number of species of vertebrate and invertebrate animals that we yank out of the oceans and lakes and rivers and fish farms knows no bounds. Their experience is surely surreal—minding their own business as they swim free in a three-dimensional volume of water. The concept of flying does not exist because if they want to ascend from their current depth, they just swim there. That's their entire world. The only existence they know. Then, all of a sudden, one gets yanked from above and pulled into a parallel universe. Nothing is familiar. The sky, the clouds, the Sun's warmth beating down. The water's surface was the edge of their oceanic universe—their cosmic horizon. They've never before seen it from the outside. Only from within. Moments later they begin to suffocate, and after they're tossed into a pile of crushed ice, they freeze to death. Those are the lucky ones. The unlucky ones get thrown back into the ocean where they try hard to convince their fish friends of what they experienced. Just another fish tale of an alien abduction.

The efficiency of meat production in the US and around the world unfolds at the expense of animal happiness and dignity, typically without regard to their pain and suffering. An entirely traceable posture in the West, given our ego and the widespread influence of this verse in Genesis:[10]

> *And God said, Let us make man in our image, after our likeness: and let them have dominion over the fish of the sea, and over the fowl of the air, and over the cattle, and over all the earth, and over every creeping thing that creepeth upon the earth.*

With rare exceptions,[11] in which a form of vegetarian theology recasts the word "dominion" as "stewardship,"[12] this passage has for millennia offered divine carte blanche for humans to do what we please with all other animals on Earth—land, sea, and air. Since the 1970s, however, the ethics of animal treatment has birthed an entire subfield of academic philosophy[13] and has become a subject of persistent activism.[14] Even if you cared nothing for the environment you could easily justify not eating meat on these grounds alone.

As we say in mathematics, there are separable variables in that argument. Suppose all animals consumed by humans were humanely raised and treated. Suppose further that they led full lives and were slaughtered without pain. That might bring some people back from vegetarian-land, especially when you consider that killing and eating animals is not the unique province of humans. Entire branches of the animal kingdom are pure carnivores: lions don't long for kale salads while mauling zebras; snakes don't forage for berries; owls don't ogle the broccoli in your garden.

If we value sentience, then we could rank animals by the complexity of their nervous system, and either eat none of them, or invoke some kind of cutoff. Mollusks okay? How about shellfish? Regular fish okay? Maybe not mammals. We're mammals. Mammals have big brains and suckle their young. How about insects? Quite the source of protein, I'm told. Ever seen one under a microscope? Low-power will do. The level of detail and function of all its body parts is staggering. Yes, they have brains too. More legs than we do, many also fly. They also know full well how to communicate with others of their own species. On top of that, most times when you peek in on them, they're going somewhere fast or doing something that looks important.

Speaking of mollusks, in the 1970s Ingrid Newkirk, a resident of Maryland, bought live snails one evening[15] upon hearing how easy they are to cook for dinner. A base preparation requires some garlic and white wine, which she had on hand, and voilà, you've got escargots. While she was driving home, the paper bag that contained the snails had unfurled on the passenger seat of her car and was open to the light. Snails have poor vision, but they can see light, which attracts them.[16] After a while Ingrid looked down and noticed the snails had crawled up to the rim of the bag and were all in a row, looking at her with their innocent, sad, beady eyeballs atop the pair of meandering tentacles on their heads. In that moment Ingrid stopped the car, released them back to the wild, and never ate snails again. In 1980, Ingrid Newkirk would cofound PETA (People for the Ethical Treatment of Animals), the largest animal welfare organization in the world. So at least for some people, mollusks: not okay.

I've seen all manner of justifications to eat or not eat one branch of animals versus others in the tree of life. Consider also the huge outcry against net-caught tuna because an occasional air-breathing mammalian dolphin gets caught in the net,[17] suffocating because it can't swim back to the surface and catch a breath. Tragic, indeed. Personally, I seek out line-caught tuna for that reason. Amid the outpouring of sympathy for the dead dolphin and all the lobbying to save them, where's our collective concern for the dead tuna? We have none. Because they're destined for the sushi bar or small tin cans on your supermarket shelves. Imagine if a delicatessen started offering dolphin-salad sandwiches for its lunch offerings. Protests would surely amass out front. Not because the deli serves sandwiches that contain dead chickens, dead turkeys, dead pigs, dead cattle, dead salmon, and yes, dead tuna, but because it serves dead dolphins.

The urge to treat one species of animal differently in any way is called speciesism. Think racism or sexism, but in this case, you're biased against some animals simply because of their genetic distance from humans in the tree of life, or because they're repulsive to look at. How many animal lovers march with placards demanding to save the leeches, mosquitoes, ticks, tapeworms, and lice? Or how about the guinea worm, whose primary host is humans? We'd just as soon see them all go extinct. Hardly anybody makes plush toys of these parasites, yet they're all creatures on God's green Earth, just trying to survive like everybody else. Can't blame them for not being cuddly mammals with adorable eyes and bushy tails.

Following this argument further, one could choose not to eat animals at all, living life as a vegetarian, but when you think about it, that's being speciesist against plant life. For example, maybe you live in a plush suburb. You capture a mouse in your basement using a humane trap. You then re-release the mouse into the wild. You feel good because you're against killing animals. What you've done, however, is add to the tray of tasty snacks for owls, hawks, snakes, foxes, and other vertebrate predators, unwittingly sealing an early death for the hapless mouse. Its life expectancy is far greater living in the warmth and safety of your home.[18] Meanwhile, your home is constructed from up to fifty full-grown trees,[19] each having lived a half century.[20] They were all cut down and milled to make the studs that framed your house, the structural beams that support it, and the hardwood floors upon which you walk. That's 250 tons of what was once oxygen-producing plant life.[21] A chubby mouse weighs an ounce. In a single day, each tree produced up to fifteen times the mouse's mass in life-sustaining oxygen.

What does nature itself care more about, the mouse or the tree? When you cut a tree, does it not bleed? (What is authentic pancake syrup, if not 30 times concentrated maple tree blood?[22]) When you enshroud a tree, does it not suffocate? When you deny water and nutrients to a tree, does it not wither and die?

What if brainless plants were secretly sentient? That concept might be hard to embrace because we're biased by brain chauvinism. Modern computer scientists face similar challenges as they assess whether robots programmed by humans can ever be sentient. Sifting the science from the pseudoscience in the research on plant consciousness, we now know that a communication network of electrochemical signals connects microbes, low-lying plants, animals, and trees. It thrives underfoot in a fungal root system of the forest called the mycelium.[23] Many think of it as a Wood Wide Web. The behaviors expressed by the participating lifeforms have been analogized by botanists to various human emotional states such as pain, joy, fear, and anger.

The ecosystem of the sci-fi fantasy world in the film *Avatar* (2009) was partly inspired by these discoveries—an exoplanet teeming with interconnected plant and animal life that shares feelings and ideas. Famous sentient plant life in the world of fiction includes the creepy talking apple trees in *The Wizard of Oz*; the old, wise, and contemplative trees called Ents from *The Lord of the Rings*; and the barely literate, lovable chunk of driftwood from Planet X named Groot, best known from his appearances in the *Guardians of the Galaxy* comic book and film series. The affable alien in the 1982 film *ET: The Extraterrestrial* had a way with vegetation. On several occasions, ET would extend its glowing index finger and magically heal dying plants. A natural talent, perhaps. I have it on good authority[24] that ET was, in fact, conceived as a sentient plant and not an animal.

Those examples are all Hollywood. Let's instead conduct an extraterrestrial thought experiment. Imagine a pod of aliens come to visit who derive all their energy and nourishment from starlight and minerals. What would they think of life on Earth? They'd see their cousins—all that photosynthesizes—and delight in their taxonomic diversity, from microscopic cyanobacteria in ponds, lakes, and seas to the mighty sequoia trees of the American northwest that live for thousands of years. They would see all other life as hopelessly barbaric, killing all manner of living things for their survival. They would see humans as apex predators—persistent purveyors of violence—dividing themselves into those who kill and eat animals and those who kill and eat plants.

Even when we're playful, we're barbaric. From the 1950s through the 1990s, millions of children grew up watching on television the ventriloquist Shari Lewis converse with her adorable sock puppet, named Lamb Chop. In 1993, Lamb Chop even testified before the US Congress on the push for quality children's television.[25] Lamb Chop. A totally cute name until you think about it for five seconds. The puppet is a lamb. The puppet's name is what happens to a lamb—a juvenile sheep—when you slaughter it, rip out its tiny rib cage, and grill it. If you had a pet pig, would you name it Hambone? If you had a pet cow, would you name it Ribeye?

The subliminal message: Lamb Chop wasn't a puppet. Lamb Chop was dinner.

Morbid as that is, our visiting light-metabolizing aliens would be especially incensed by Earth vegetarians for slaughtering their plant brethren. Not only that, human plant eaters take special interest in reproductive organs—the flowers, the seeds, the nuts, the berries—and eat those, disrupting the life cycle of the plant.

Plenty of other fruit-eating mammals delight in such snacks, often swallowing the hard-cased seeds whole, which then pass through their digestive tract unscathed. By then the animal has wandered to new places where the seeds emerge, embedded in free fertilizer. The plant has passively spread its presence across the countryside through a symbiotic relationship with hungry mammals. Isn't nature beautiful? We humans, however, grind the seeds of fruits and berries to a pulp with our molars. Those we happen to swallow whole don't propagate the plant's life cycle because we (generally) don't poop in open meadows.

It doesn't end there. Barbaric humans further seek out the youngest versions of plants to harvest. Why else would the produce aisle in grocery stores contain infanticidal packages of baby carrots, baby spinach, baby arugula, baby artichokes, baby squash, bean sprouts. The list goes on.

A blunt truth of human existence: our three sources of energy—protein, carbohydrates, and fat—all come from killing and eating other forms of life in our ecosystem. We can get some of our necessary minerals, like salt, from the environment, but you can't live on minerals. Two foods rise above the "I must kill to survive" way of life: milk and honey. Combined, the two are rich in protein, carbohydrates, and fat and don't require the death of any living thing for your nourishment. If you don't otherwise metabolize sunlight, a milk-and-honey diet would be the least violent way you could possibly live on Earth.

Note that milk and honey are specifically excluded from the diet of vegans, on the grounds that you're taking food intended for calves and bees. I imagine that lactating cows and bees don't want their precious nourishment taken from them, although they can probably just make more of it. In any case, vegan philosophy prefers you kill plants for your nourishment, rather than steal milk from the cow and honey from the bee.

Given the pace of food innovations, we may soon spawn an entire cuisine of lab-grown meat. These cultivated proteins look like meat and taste like meat because . . . they are meat. The production line simply doesn't require you to raise and kill any living organism. These products can be infused with vitamins, minerals, trace nutrients, and even chef-tailored flavors that don't require seasoning at home. Many companies attempting this are even publicly traded.[26] So the marketplace is priming itself for the industry to flourish. Throw milk and honey back into that mix and if vegetarians return across the divide, we may see a future of civilization where neither plants nor animals are killed to support the lives of humans. We'd then be protected during the next visit from angry aliens who don't eat plants and angry aliens who don't eat meat. Apart from all the killing we do of one another, they might view us as the most nature-loving species in the Galaxy.

There's another comically true difference between eating plants and eating animals. The successful TV producer Chuck Lorre, perhaps best known for co-creating the hit sitcom *The Big Bang Theory*, posts what he calls "vanity cards" at the end of each episode. He uses them to briefly opine on one subject or another, and they are visible for just a second or two on the screen. They contain vastly more text than can be read in that time, so one must find them on the internet. After apologizing in advance for whom he might piss off, Lorre's card no. 536 contains the following attack:[27]

> *Vegetarians and Vegans are mobility bigots. They believe that if a life form doesn't move, it's fair game to be killed and eaten. . . . This hateful philosophy is predicated on the idea that movement equals consciousness, or, if you will, a certain level of sacredness. . . . Of course when you ask vegetarians and vegans, they say no, they're only opposed to eating flesh. But what could be more fleshy than a mushroom? Or avocado? Or eggplant? The ugly truth is they are cowards who murder and devour anything that can't run away.*

He goes on to worry about his uncle Murray, who often sits motionless for hours in front of the TV. Like plant life, Uncle Murray hardly ever moves, so he might be spotted by vegetarians and eaten. Need I remind the reader that Lorre writes successful sitcoms. So this bit of satire should be viewed as entertainment. He's so good at what he does that if you type *The Big Bang Theory* into a Google search engine, the top hits are all his TV show. You must scroll down some more before you find any discussions on the origin of the universe. As an astrophysicist-educator, I'm still trying to figure out whether that's a good thing or a bad thing.

The sanctity of animal life, briefly addressed by Chuck Lorre as a vegetarian edict, has deep roots, although the seventeenth-century Dutch polymath Christiaan Huygens went one step further. He grouped plants and animals together and offered a divine comparison between them and the rest of nature.[28]

> *I suppose no body will deny but that there's somewhat more of Contrivance, somewhat more of Miracle in the production and growth of Plants and Animals than in lifeless heaps of inanimate Bodies. . . . For the finger of God, and the Wisdom of Divine Providence, is in them much more clearly manifested than in the other.*

Maybe it's all sacred. Maybe one day, the organization PETA will encounter a rival organization, PETP (People for the Ethical Treatment of Plants). Or maybe humans are a peculiar aberration to the natural order of the universe. What are we to a grizzly or polar bear? Are we sentient beings, capable of art and philosophy and science and civilization? No. We're free-range meat. Every part of us. The cartoonist Gary Larson and his morbid sense of humor captured this brilliantly in a comic that portrays a hungry polar bear who bites open a hole in the top of an igloo and excitedly describes the meal to a fellow bear as "Crunchy on the outside and a chewy center!"

Want more? In a 1991 short story first published in *Omni* magazine titled "They're Made out of Meat," sci-fi author Terry Bisson makes you regret being human. We are treated to a conversation between two ethereal aliens, where one tries hard to explain to the other that Earth humans are made entirely out of meat. A snippet of their pithy dialogue captures the astonishment:[29]

They're made out of meat.

Meat?

Meat. They're made out of meat.

Meat?

There's no doubt about it. We picked several from different parts of the planet, took them aboard our recon vessels, probed them all the way through. They're completely meat.

That's impossible. What about the radio signals? The messages to the stars.

They use the radio waves to talk, but the signals don't come from them. The signals come from machines.

So who made the machines? That's who we want to contact.

They made the machines. That's what I'm trying to tell you. Meat made the machines.

That's ridiculous. How can meat make a machine? You're asking me to believe in sentient meat.

The first alien later attempts to describe how humans communicate:

> **You know how when you slap or flap meat it makes a noise? They talk by flapping their meat at each other. They can even sing by squirting air through their meat.**

To offer more perspective, consider that all species large and small in the tree of life are contemporary players in Earth's land, sea, and air ecosystem. The largest known organism in the world is a single mat of mushrooms weighing 35,000 tons (nearly two-thirds the weight of the RMS *Titanic*). This humongous fungus lurks underground and measures miles across in the Blue Mountains of Oregon. If you're into hard-to-pronounce, hard-to-remember italicized names for genus and species, it's called *Armillaria ostoyae*. Mushrooms occupy their own kingdom of life, which split with animals in evolutionary history later than our common ancestor split from green plants. Humans and mushrooms are therefore more genetically alike than either we or mushrooms are to anything that grows in the plant kingdom.

Maybe that's why we commonly say mushrooms taste "meaty"—an adjective never offered to kale. In a distant ancestral way, we are biting into ourselves.

SEVEN

GENDER & IDENTITY

People are more the same than different

The lines of division in modern civilization seem endless. We willingly sort ourselves by hair color, skin color, what we eat, what we wear, who we worship, who we sleep with, what language we speak, what side of the border we live on, and so forth. In the universe, astrophysicists are no strangers to this exercise. Matter and energy express themselves across a staggering breadth of properties that include measures of size, temperature, density, location, speed, and rotation. In some cases, nature divides cleanly into categories that we can define unambiguously, like whether a substance is a solid, liquid, or gas. You've probably never in your life been confused about which is which.

Although even those distinctions have issues.

You may have heard, even if not a resident of the Mountain Time Zone in the US, that at high elevation you must increase cooking times to compensate for the lower air pressure there. But hardly anybody ever tells you why. The boiling temperature of a liquid is not some universal constant. It depends on the air pressure pushing down on the liquid's surface. If you reduce the air pressure on water, it will boil at a lower temperature, forcing you to cook food longer just to compensate. Keep lowering the air pressure and you will keep lowering the boiling point. If you lower the air pressure well below the level where you'd suffocate and die, there exists a pressure and temperature for which water boils as it freezes. Under those magic conditions, the solid, liquid, and gaseous forms of water all happily coexist in what is sensibly called the triple point of water. Large swaths of the Martian surface happen to satisfy these conditions. So the question, "What is the state of water at the triple point? Is it solid, liquid, or gas?" has a simple answer: "Yes." All three, simultaneously. An odd but accurate reply that's completely sensible if you loosen the urge to compartmentalize everything around you.

To require that objects, things, and ideas fit into neat categories apparently runs deep and derives from an inability to cope with ambiguity. Are you with us or against us? Maybe the answer is somewhere in between—or everywhere in between. We fight against it with all our might.

The perplexing wave-particle duality of matter disturbs many people. The term "wavicle" never caught on. Maybe it should have. The wailing takes the form: "Which is it? It's gotta be one or the other. I must know!" The simple answer is that matter manifests as both waves and particles. Get over it.

Is the infamous Schrödinger's cat[1] dead or alive in the closed box? If you open the box, you will discover the cat to be either dead or alive. Yet quantum physics tells us that if you don't open the box, the cat is both dead and alive at the same time. Get over that one too. Nature carries no obligation to accommodate our limited capacity to interpret reality. A cat in a box is just the beginning. As you read this, quantum computing is being invented. A new kind of circuitry that embraces the statistical uncertainties and binary ambiguities of real life problems in this world. In classical computing, all calculations—and all data—exploit whether a "bit" has the value of 0 or 1. Yes, it's all 0s and 1s. Our infotech universe is binary.

Quantum computing instead uses "qubits." A qubit can be a 0 or 1, just like its classical cousin. But a qubit can also be a continuous combination of 0 or 1: a little bit of 0 and a lot of 1; a lot of 0 and a little bit of 1; equal amounts of both; and everything in between. In quantum-speak we call this a superposition of the two states. Not knowing whether a qubit is a 0 or 1 is not a shortcoming of quantum computing; it's a coveted feature that challenges our binary brains to embrace it.

In the universe, two or more seemingly contradictory facts can be simultaneously true. How about on Earth? Can you be both male and female? Can you be neither? Can you move fluidly between being a man and a woman? Is your sexual preference fluid too? Maybe we're all male-female qubits. Such questions are hard for some people to grasp, embedded in a culture that sees the world as a landscape of rigid categories, where things must be one or the other, and not fall on a continuum.

An analysis of colors offers insight. For simplicity and convenience, we speak of the seven colors of the rainbow—the seven colors of the visible solar spectrum: red, orange, yellow, green, blue, indigo, and violet. You can remember the sequence because their initial letters make the acronym for somebody named "Roy G. Biv." We embrace these seven colors, occasionally omitting the undeserving indigo—leaving six—as in the modern, stylized NBC peacock and in the traditional LGBT flag. What hardly anybody talks about, but which astrophysicists know deeply, is that the colors from red through violet fall on a continuum. If we possessed the visual acuity and attendant vocabulary to describe them, we could identify thousands of colors and their hues, blending seamlessly. No sharp boundaries to be found anywhere. The colors of light form a continuous sequence of wavelengths, which also tracks energy and frequency. When astrophysicists talk about an object's color, we can do so with high precision, referencing specific wavelengths of light without invoking coarse color categories that are commonly used.

The list of letters currently represented by the rainbow flag is LGBTQ+: lesbian, gay, bisexual, transgender, queer, and others with nonconforming gender and sexual identities. Of those words, "gay" and "queer" were at one time pejorative. The community and the associated movement reclaimed them, removing the power that oppressors wielded when armed with those words on their tongues. At last count, there are at least seventeen nonconforming designations,[2] each identifying fellow humans who are not cisgender heterosexual, itself referencing a person whose inner identity and gender correspond with their assigned sex and whose mating preference is the "opposite" sex. That cis-het image is basically all we ever saw in the storytelling of movies and television for most of the twentieth century. Those who didn't fit the mold weren't simply other characters that happened to be in the story. They were instead singled out for comedic, verbal, or physical abuse. In the 1961 version of the film *West Side Story*,[3] a kid named Anybodys wants to be part of the Jets gang of boys. She's got short hair. She has a dirty face. She's spunky. She's ready to fight. She wears pants. There's nothing dainty about this archetypal Tomboy. No, they won't let her join the Jets because . . . she's a girl. If you're not a boy, you're a girl. A brief exchange with Riff, the Jets gang leader, along with A-Rab and other gang members conveys the rejection:

Riff: Not you, Anybodys. Beat it.

Anybodys: Aww, Riff, you gotta let me in the gang. . . . I'm a killer. I wanna fight.

A-Rab: How else is she gonna get a guy to touch her?

Riff: Come on, the road, little girl! The road!

Jets Gang: Beat it!

The world was quite binary back then, even though we all knew and saw people around us, if not ourselves, who did not conform.

This boy-girl thing runs deep. You might have guessed that the Bible has a verse about it:[4]

The woman shall not wear that which pertaineth unto a man, neither shall a man put on a woman's garment: for all that do so are abomination unto the LORD thy God.

Clearly, the Creator of the universe cares about your choice of wardrobe. Joan of Arc, a kindred spirit of Anybodys, was convicted in 1431 during a trial that famously cited her persistent cross-dressing as one of the heretical offenses that led to her being burned at the stake.[5]

The urge to categorize and create an “other” is strong, perhaps because it’s more difficult to think about whether a continuum connects you to that person, or to any other person you deem different from yourself. Biology itself won’t bail you out of this one. The presumed binary of sex in nature is overrated and rife with exceptions, not only in ourselves but also in the rest of the animal kingdom.[6]

What then of our physiology? What of our chromosomes—the famous XX and XY pairings that clearly and cleanly designate female and male identity in most humans? When you decide who is male and who is female what do you cue on? Here's an experiment anyone can conduct: One cold winter’s day on the NYC subway I observed everyone who was seated—a typical diverse bunch just going to work in the morning. We were all wearing puffy, warm, dark outer garments, so no one’s body shape was discernible beneath their coats. All you could see of anyone was their head. Note also that the length of our legs carries nearly all height difference between humans. When seated we are all approximately the same height, which is why driver car seats adjust forward and backward with vastly more range than they adjust up and down, if at all. I then gave myself a gender test. Could I identify who presented as male and who presented as female just from their faces? It was easy. Even after removing from the sample those obvious man-heads with facial hair and male-pattern baldness. What I cued on was societal norms for what males and females should look like from the neck up. For that sample, the data were binary. The women, on average, sported longer hair. They were more likely to wear earrings, and if they did, the earrings were larger. Their eyebrows were tweezed. They were more likely to wear visible makeup like eyeliner, mascara, rouge, and lipstick. They were more likely to wear visible jewelry like necklaces, expressive finger rings, and bracelets. Their hands were more likely to have painted fingernails.

Of course some men displayed some of these features. But weighing all factors, it was clear who the boys and who the girls were. That's when I realized: 100 percent of my cueing was on secondary and tertiary features—all societal constructs. I then carefully imagined everyone on the subway without such adornments. A hard but not impossible task. Upon doing so, I could not distinguish one gender from another. Was there a canonical female-shaped face or male-shaped face that would reveal their gender to me? The nose, forehead, cheekbones, jawline, lips, mouth? No trends to be found. Reading up on related literature, I found comments that men possess a distinct masculine jaw and brow relative to women.[7] With attendant notes like, if a woman has these features, then she simply has a masculine face, and if a man has soft facial features, he's feminine. They could have instead declared that the range of all people's faces includes both soft and distinct features, without genderizing them.

That's when I reckoned the massive investments we all make in gender expression. Want to look more like a man? Maybe grow a mustache or beard. But certainly visit the gym and build some muscles. Buy clothes only in the men's section of department stores. Fashion designers have already thought about this for you. Choose clothes that reveal your new physique. Want to look more like a woman? Remove hair from above your lip, between your eyebrows, up and down your legs, and other places you might designate. Because everyone knows that men are hairy and women are not. Breasts not large enough? Those are important because men don't have breasts and women do. So why not boost those. Wear bras that accentuate what you have, or surgically increase their size, as more than 200,000 women do annually in the US.[8] Buy clothes only from the women's section of the department store. They definitely know what you're supposed to look like.

Without these tools and social standards available to us and without our daily investment in gender expression, how different would we really look from one another? How androgynous would we all become? With or without an overcoat? Believe it or not, Santa's reindeer exemplify the problem. Unlike other deer species, both male and female reindeer grow antlers. So at a glance they all look the same. But zoologically all male reindeer lose their antlers in the late fall, well before Christmas.[9] In spite of their names, only some of which are feminine,[10] all Santa's reindeer sport antlers. So they're all female. Which means Rudolph has been misgendered.

We tend to categorize information even when granted full prior awareness that important features are continuous and do not lend themselves to simple sorting. Consider the Saffir-Simpson scale, which divides hurricanes into five categories.[11] Not three. Not nine. Not twenty-two. With sharp boundaries between them, measured by sustained wind speed.

Category 1	74 to 95 miles per hour
Category 2	96 to 110 miles per hour
Category 3	111 to 129 miles per hour
Category 4	130 to 156 miles per hour
Category 5	exceed 157 miles per hour

Anything magical about these boundaries? Not entirely. They're not even round numbers when converted to metric. Herbert Saffir, a structural engineer, and Robert Simpson, a meteorologist, implemented the scale in 1973. They correlated wind speed with structural damage sustained by homes and buildings of the time. To this day, weather forecasters wait with high anticipation for when a hurricane's intensity grows (or shrinks) from one category to the next. That upgrade counts as breaking news. Meteorologists will duly report, "Hurricane Hilda has strengthened from Cat 3 to Cat 4." But will rarely say, "Hurricane Hilda has strengthened from low Cat 3 to high Cat 3." There's more of a difference between a weak and strong Category 3 hurricane than between a strong Category 3 and a weak Category 4. Yet this distinction is lost because we've tossed hurricane strengths into only five bins. No harm here, other than serving as more evidence that our brains don't do well on a continuum, preferring instead to manufacture categories that nature itself does not provide.

How many categories is the right number? Is that even the right question?

The universe is vast and varied, daily forcing scientists to confront, accommodate, measure, and analyze the diversity of all that's out there. As an astrophysicist, I easily embraced the six-color rainbow flag as a full-spectrum emblem of all people. New versions of the flag sport even more colors,[12] calling explicit attention to additional, previously unrecognized, nonconforming groups. This transforms the flag from a symbol of the continuous spectrum to the representation of discrete groups. One day, we may discover or otherwise affirm no discrete categories at all, as the multidimensional gender universe unfolds along a continuum, like the colors contained in sunlight. This will significantly dilute the power of homophobic and transphobic bigots to declare that they are somehow separate and distinct from other members of their own species.

Many people, who defend our cherished freedoms as citizens of the US, will argue against mandated masks, helmet laws, gun laws, seatbelts, and against anything else that constricts a person from living life the way they want. Odd that many of these same people will maintain or seek laws to restrict another person's free expression of their gender identity.[13] Yet somewhere I read that life and liberty were foundational to America—a seminal experiment, begun in 1776, in how to be a country. Somewhere I read that the pursuit of happiness is something worth fighting for. Somewhere I read that the United States was the Land of the Free. None of this can be true unless the players in this experiment embrace rational thinking by rising above themselves and looking back to see the hypocrisies that infect their thoughts and decisions. Imagine how free the world would be if we all took a "hypocritic" oath:

I shall never claim to have moral standards or beliefs to which my own behavior does not conform.

EIGHT

COLOR & RACE

Once again, people are more the same than different

In the early twentieth century, we figured out that stars themselves can look very different from one another when you analyze their spectra. At the time, a roomful of computers classified tens of thousands of stars by their "spectral type," binning them into fifteen lettered categories: A through O. With better data and with an emergent understanding of quantum physics, these categories would be culled and further divided into ten numbered subcategories as well as nine other categories represented mostly by Roman numerals that track a star's evolutionary state. In recent decades, several additional spectral types would accommodate very dim stars that were not yet discovered when the original data were compiled. A coded system of three dozen additional symbols would further accommodate highly unusual or peculiar features on top of an otherwise ordinary star.

That roomful of computers was composed of sentient carbon-based life—computationally fluent and scientifically literate women hired by the men of the Harvard College Observatory to do the tedious measurements and bookkeeping of all the data.[1] Little did the men know that spectral classification was not just the act of cataloging data, it would establish the entire subfield of stellar evolution. Sexist culture notwithstanding, we are aware that stars in the Galaxy and across the universe represent a continuum of properties that we slice into hundreds of bins for our conversational and scientific convenience. The Sun, if you are curious, is spectral class G2V. Polaris, the North Star, F7I.

Back on Earth, many places will label your skin color black or white or brown. That's it. When you report to police whom you glimpsed committing a crime, they fully expect you to draw from only one of those three categories. The world's staggering range of human skin color has been reduced to three shades, and somehow everybody's okay with that. Maybe you added red and yellow, to include Native Americans and Asians, even though nobody's skin is any of these colors. White people don't disappear when they walk in front of snowbanks. And while skin color can get very dark, nobody is pure black. You've surely never met anyone fire-engine red or lemon yellow. So our color categories are simply lazy and feed whatever racist proclivities we might otherwise possess. Consider that President Barack Obama's mother was White American with European ancestry. His Kenyan-born father was, of course, African. Obama arrived in this world with a skin color halfway between the two—a light-skinned Black person. In the US, Obama was America's first Black president. Now imagine Obama as the light-skinned leader of an African country. If we invoke symmetric reasoning, that population could justifiably see him as their first White president.

In astrophysics we have a word for the reflectivity of a surface: "albedo." We invoke it often when analyzing how much solar energy a planet's surface absorbs compared with what gets reflected by cloud tops or shiny topography. A surface with an albedo of 0 absorbs all incoming energy. An albedo of 1 reflects it all away. Earth's albedo, averaged over all regions and all seasons, checks in at about 0.3, which means we reflect back 30 percent of the Sun's energy and absorb 70 percent. The energy a planet absorbs drives climate. Some people want to solve global warming, not by reducing their greenhouse gas footprint, but by injecting reflective particles high in the stratosphere.[2] This will increase Earth's albedo and reduce the amount of sunlight available for Earth to absorb.

If you really wanted to document the lightness or darkness of people's skin, you could measure everyone's albedo. Doing so would quantitatively reveal the obvious fact that a continuum of reflectivity manifests among the world's humans. A study of deeply indigenous populations of the world, communities that have lived in their locations for thousands of years, reveals a strong correspondence between latitude on Earth and skin tone.[3] The closer to the equator, where sunlight is most intense, the darker your skin tends to be. The farther from the equator, as you near the poles, the lighter your skin. By this measure, if Santa Claus is indigenous to the North Pole, he would be the whitest person there ever was. The occasional efforts to portray a Black Santa, while motivated by the noble quest of inclusion, simply ring untrue. Same reasoning applies to the imagery of a white Jesus. Having hailed from sunny Nazareth (latitude 33° North), Jesus would likely have been several shades darker than he is persistently portrayed in Renaissance frescoes and Hollywood films.

The indigenous skin color map of the world correlates with how much harmful ultraviolet light from the Sun reaches Earth's surface at that latitude.[4] We know that the molecular pigment melanin—the active ingredient of dark skin—can dissipate 99.9 percent of UV light. We also know that UV can destroy skin cells in ways that lead to sunburn and eventually cancer. Skin pigments arose as a marvelously adaptive feature of human evolution that can, in fact, be achieved through several different genomic pathways.[5] Yet people persist in sorting our species into just a handful of colors.

Odd, because the hair-products aisle in your local pharmacy displays no fewer than a hundred hair colors, each with unique names, such as cinnaberry, cinnamon stick, brass, hot cocoa, nutmeg, Brazilian dusk, peppercorn, roasted chestnut, Sahara blonde, and coastal dune blonde. That's just a subset of one line of hair color within a brand.[6] Further, the makeup aisle tries hard to match your skin color precisely. Since most makeup is intended to blend with your natural complexion, cosmetic companies are forced to think about skin on a continuum. The professional makeup artist—a literal artist—will mix a range of base colors to match your own. Think of it as a superposition of states, to borrow quantum language. The makeup products don't typically reference words but numbers or codes, which lend themselves to much more precision and nuance, just like the spectral classification of stars. From the makeup sessions that precede my appearances on television, I happen to know that in the MAC line of cosmetics, the concealer that comes closest to my facial skin tone is NW43. Beyond makeup, the actual kings of color are interior decorators. Wondering what color to paint your walls? Benjamin Moore lists thousands upon thousands of hues[7] from which to choose, not to mention a hundred shades each of white and black—all with inventive names accompanied by a numerical code. Further evidence that if we wanted to, and tried real hard, we could embrace more than a few color categories to describe people's skin to the police.

Why classify by skin color at all, unless you plan to invoke it in some way. If one group oppresses another, inadvertently or on purpose, you'd want good data on who's oppressing whom so that you can redress the problem. With the same data, however, nefarious people in power might want to magnify inequality, which is just what happened in apartheid South Africa. The 1950 Population Registration Act codified skin color into White and Black, with multiple subcategories of Coloured, which included mixed race and Asian, enabling the White minority in power to establish laws that prescribed and stratified the social, political, educational, and economic freedoms of each population differently.

Is out-group hatred really just about having a different skin color? Apparently not. Both the First and the Second World Wars in Europe were mostly very light-skinned people killing other very light-skinned people, reaping a combined death toll of more than eighty million. Looks like differences in language, ethnicity, politics, and cultural values fed the conflicts more than skin color. Where does it stop? The thirty years of bloodshed within Northern Ireland, politely called the Troubles, was basically a political fight for control of the region. The warring factions happened to divide between Irish Catholics and Irish Protestants. But when viewed from afar, either ethnically, geographically, nationally, or stratospherically, it was White Christians finding reasons to slaughter other White Christians. More than 3,500 people died over that time—a stark reminder that while skin color can make a person an easy target of hatred, it's not a prerequisite for wanting to kill.

The prevailing movement in the US and elsewhere to purge public statues of people with racist pasts reached a fever pitch after the May 25, 2020, murder of George Floyd by a Minneapolis police officer. In 2020 more than one hundred monuments were removed, mostly of Confederate Civil War commanders—all in full uniform, many on horses. Statues such as these, especially those that stood silent guard over Monument Avenue in Richmond, Virginia, all remind me of slavery. We've come a long way since the Civil War. I've even been awarded an honorary doctorate from the University of Richmond. During my life, I've achieved what was surely unthinkable to each of these Confederate leaders—and honored for it in the cradle of the American South and birthplace of General Robert E. Lee.

Yes, that's progress, but these statues nonetheless evoke in me the country's darkest era. I'm not debilitated by these thoughts, but they do represent a kind of socio-emotional tax I pay. Maybe all these men were kind and noble members of their community. Maybe they were pious Christians and went to church each Sunday. Maybe they helped rescue cats from trees. Could all be true. That's not why these equestrian statues were cast in their honor. The soldiers are memorialized for defending the Southern way of life from an incursion of Northern aggressors. But there is no Southern way of life that isn't economically and culturally conjoined with the fate of four million enslaved Africans—fully a third of the South's population.[8] To sustain the hierarchy requires an endemic belief system of racial inferiority, complete with laws that prevented the education of Negroes. A key step when you want to feel superior: can't have stupid, inferior Black people walking around who are more educated than you are.

The few sentences below say it all, spoken in 1836 during a two-hour-long oration to Congress, in defense of slavery, by South Carolina representative James Henry Hammond. To him, slavery was:[9]

> *. . . the greatest of all the great blessings which a kind Providence has bestowed upon our glorious region. For without [slavery], our fertile soil and our fructifying climate would have been given to us in vain. As it is, the history of the short period during which we have enjoyed it has rendered our Southern country proverbial for its wealth, its genius, its manners.*

This knowledge prevents me from seeing Confederate military statues as quaint reminders of a bucolic, agrarian past. Nor do they engender sympathy for a noble, lost cause. Without the African slave trade there are no romanticized plantations to feed the South's rose-colored memory of itself.

Mixed within this movement to dismantle or relocate statues was the successful call for the American Museum of Natural History[10] to remove the massive statue by James

Earle Fraser of Theodore Roosevelt parked in front of the main entrance. Roosevelt sequentially held the offices of New York City police commissioner, New York State governor, US vice president, and of course, president, serving from 1901 to 1909. After the Statue of Liberty, the museum's Roosevelt statue is one of the largest in the entire City of New York. He's astride a majestic horse with one hoof raised—equestrian code to indicate he served in the military. Though he was a US Army colonel, he's dressed not in uniform nor in a president's suit and vest, but in casual clothes with sleeves rolled up. So why the controversy? Could it be things he said? Here's a memorable quote, speaking to the New York Republican Club in 1905, on the subject of those unfortunate enough to be born with dark skin:[11]

> *The problem is so to adjust the relations between two races of different ethnic type that the rights of neither be abridged nor jeoparded; that the backward race be trained so that it may enter into the possession of true freedom while the forward race is enabled to preserve unharmed the high civilization wrought out by its forefathers.*

Remember that the Republicans back then held relatively progressive views of society and the world. Here, Roosevelt does not want to re-enslave Black people. He wants them to participate in the American dream, to whatever level their backward morality and intellect would allow.

Roosevelt was a learned man. While he surely could have fostered such thoughts on his own, these ideas were deeply infused by the prevailing scholarship of eugenics—the academic study of how to create a better race of humans by breeding good features and suppressing bad ones. How? Just discourage undesirable people from making babies, or in the case of a "feeble-minded" diagnosis by a eugenicist, sterilize them to prevent breeding at all. This branch of social biology—with leading advocates at prestigious institutions such as Harvard and even the American Museum of Natural History—influenced politics, laws, immigration rules, and the social order of America for decades.[12] The movement even provided buoyancy for the social policies of Nazi Germany. Since Roosevelt was of his time, I'm going to give Teddy a hall pass and cast much, if not all, of the blame on my fellow scientists from another field—from another generation.

In any case, Roosevelt's quotes on race are not what inspired the statue. Apart from having held high office, he wrote extensively on statesmanship, conservation, exploration, and the value of wildlife. During his presidency, he set aside 230 million acres of public land, birthing the United States Forest Service. These principles and others are why he is the patron saint of the American Museum of Natural History, with a huge rotunda at the main entrance lined with stirring quotes from him—none of them extolling White superiority.

The statue also memorializes two other people. A Black African and a Native American, standing stoically in native garb—one on each side of Roosevelt and his horse. Seen with modern eyes and sensibilities, this portrayal is an abomination. Nobody today would dream of casting a White man on a horse flanked by oppressed, disenfranchised people. What could they possibly have been thinking?

No need to wonder. We know exactly what was going on in the sculptor's mind and in society at the time. It was commissioned in 1925 and unveiled in 1940, so consider a few facts:

- The Black African and the Native American both stand proud and distinguished in their posture and facial expression. They are both muscled and almost regal.

- The Black African and the Native American look forward, out to a distant place. The same direction as Roosevelt.

- At the time, from the 1920s through the 1940s, nearly all other representations of Negroes and Indians were embarrassing caricatures—in books and films and comics and cartoons—intended to make White people laugh and feel superior.

- At the time, hardly any statues of Black people or Native Americans were erected anywhere, apart from brass diminutive Black jockeys outside exclusive clubs and restaurants and wooden Indians rolled to the front of smoke shops during business hours.

- Ever seen any other statue of a US president with random people in it? They're rare. Three come to mind: (1) The 1997 Franklin Delano Roosevelt Memorial in Washington, DC, where FDR is portrayed with

his dog, a Scottish terrier named Fala, in a sculpture by Neil Estern. If we exclude dogs as people, then (2) FDR again, in a 1995 sculpture by Lawrence Holofcener titled *Allies*, seated adjacent to Churchill on a London bench along Bond Street,[13] although Winston is hardly a random other person. Last (3) an 1876 sculpture by Thomas Ball in Lincoln Park, in Washington, DC, titled *Emancipation*, in which Abraham Lincoln's left hand outstretches Jesus-style, over the head of a kneeling, shackled, enslaved African. Might have seemed like the right thing to erect following the American Civil War, but through modern eyes, it's embarrassingly patronizing.[14]

- In 1940, if you viewed Negroes and Indians as inferior to yourself and if Teddy Roosevelt was a hero of yours, you'd surely see the sculptor's two standing figures as an inexcusable act of denigration. How dare he sully Roosevelt's reputation with these portrayals of lesser humans.

- John Russell Pope, the architect of the new museum building behind Roosevelt, described the three men of the sculpture as a "heroic group."

- How about what James Earle Fraser, the sculptor, said himself:[15]

> *The two figures at [Roosevelt's] side are guides symbolizing the continents of Africa and America, and if you choose may stand for Roosevelt's friendliness to all races.*

As a scientist and as an educator, I care less for opinions than I do for one's capacity to think reasonably and rationally about all the relevant data that may inform those opinions. Remembering too that our opinions take shape on an ever-shifting landscape of social and cultural mores. A full perspective on the statue leaves me happy to see it removed, only because nobody would conceive of such a statue today, which is the litmus test for any indictment I seek. The statue will live in Medora, North Dakota, the site of Teddy Roosevelt's new presidential library.[16] My urge is to gesture a forgiving nod to the past while I display a furrowed brow to the future, wondering how today's most progressive creations may appear to our ever-more-enlightened descendants a hundred years from now.

Why would anyone want to feel superior to others? Surely the only occasion to justify looking down on someone is while you are helping them up. Not being a psychologist, I claim no special insight other than to share the observation itself. What's clear is that some people feel better when they believe other people are less than they are, in any way they value, which could include wealth, intelligence, talent, beauty, or education. Add strength, speed, grace, agility, and endurance and you've compiled most of the ways people persistently compare themselves to others, either informally or in organized settings. The Olympics owes its existence to the search for people who perform faster, higher, and stronger among us. Standardized exams, game shows, beauty contests, talent auditions, and the Forbes 400 all pit humans against humans, in rank order. Society offers hundreds, if not thousands of ways to show you're better than others.

What happens when your sense of superiority applies not to an individual whom you just beat at, say, a game of chess, but to an entire demographic? Most of those people you have never met and will never meet. You feel superior because somebody told you it was okay to feel that way—your parents, or some political or social authority. You might expect cultural bias to get handed down from one generation to the next, or nationalistic delusions to override rationalistic thought. You might also be convinced by a spokesperson for God that your religion is better than everybody else's. But what happens if a bona fide scientist tells you that you're superior? Will you accept the conclusion, riding high on the feeling, or will you explore possible sources of bias?

As you might suspect, among branches of scientific inquiry, those most susceptible to human bias are fields that study and judge the appearance, conduct, and habits of other humans. Topping that list, find psychology, sociology, and especially anthropology. If they are to establish and preserve their integrity, these fields must engage extra levels of peer review and disclosure, with the express purpose of spotting bias.

Because humans are not a topic of study within mathematics and the physical sciences, these fields tend to resist this blight. That doesn't mean the researchers themselves can't be racist, sexist misanthropes. Nor does it mean these fields are without bias. It just means cosmic discovery, and what lands in the textbooks, is less susceptible to feelings of superiority.[17]

The closest thing I could find to a non-inclusive concept in my math and physics training is the "hairy ball" theorem[18] from algebraic topology, proved in 1885 by the French mathematician and theoretical physicist Henri Poincaré. It enjoys many colloquial rephrasings, such as "you can't comb the hair on a bowling ball." Translated with more precision: if you combed a hairy sphere, no matter how you comb it, at least one spot will remain where the hair does not know which way to lean—there will always be at least one cowlick on the ball.

The theorem is true, but the reference to hair combing is innocently biased. I comb the hair on my bowling-ball head every day. I do so with an Afro pick, which draws the hair shaft straight away from my scalp. If I didn't have a face or neck, and my head were just a sphere, I could easily comb my hair without the anxiety of a single cowlick sticking out. It's all cowlick. If Poincaré and all others had sported an Afro, no doubt he still would have proved the theorem, perhaps minus any reference to hairstyles.

In another example, my wife, who has a PhD in mathematical physics, was quick to note that many cosmologists clutched to the idea that we may live in a steady-state universe, long after data from leading telescopes made it clear we do not. At the time, we learned that our expanding universe, birthed from a Big Bang, may one day recollapse and perhaps cycle endlessly. She wondered whether the steady-state cosmologists, most of whom have never menstruated, had a hard time thinking about and embracing cycles—something half the world's population lives with for most of their adult lives.

For one of my college electromagnetism classes from the late 1970s, when learning the formula relating charge (**Q**), voltage (**V**), and capacitance (**C**) in electric circuits, the graduate teaching assistant, with memory still green from Vietnam, offered a simple mnemonic to remember the equation: **Q** = **V** × **C**. "Our **Q**uarry is the **V**iet**C**ong." Sadly, I've remembered this ever since.

Taking that up a notch, the resistor is a common circuit element that's rainbow-coded for its resistance value in units of "ohms." (Why they never printed the numerical resistance on the resistor itself has puzzled me to this day.) There they are. Stripes of colors: Black-Brown-Red-Orange-Yellow-Green-Blue-Violet-Gray-White. Different combinations of these colors indicate different resistor values. Nothing could be more ripe for being mnemonicized than that. Wikipedia lists dozens of them in an exclusive entry on just this topic.[19] For example: **B**ig **B**eautiful **R**oses **O**ccupy **Y**our **G**arden **B**ut **V**iolets **G**row **W**ild is simple and conjures beautiful imagery. But that's not the mnemonic I first heard. In my electronics physics laboratory, I was instead taught: **B**lack **B**oys **R**ape **O**ur **Y**oung **G**irls **B**ut **V**iolet **G**ives **W**illingly. Upon spotting me, realizing I'm the only Black student in the room, the instructor sheepishly shrugged and quickly altered the mnemonic to begin with "**B**ad **B**oys . . ." thus changing a racist, misogynous mnemonic into just a misogynous one. Thankfully, none of this affects what electrons do in their circuits.

Arguably, the writings of nineteenth-century anthropologists reflect the most racist era in the history of science, spilling into the twentieth century. A fascination with the "races of man" preoccupied the lifeworks of many researchers. My favorite example comes from the 1870 study *Hereditary Genius: An Inquiry into Its Laws and Consequences*, by the English polymath Francis Galton, founder of the eugenics movement, among other branches of experimental inquiry. In the chapter titled "The Comparative Worth of Different Races," he notes:[20]

> *The number among the negroes of those whom we should call half-witted men, is very large. Every book alluding to negro servants in America is full of instances. I was myself much impressed by this fact during my travels in Africa. The mistakes the negroes made in their own matters were so childish, stupid, and simpleton-like, as frequently to make me ashamed of my own species.*

Galton was knighted in 1909 by Edward VII.

But the exercise began earlier. Just divide all humans by skin color, hair texture, and facial features, and you're halfway there. Equipped with starter data such as this, the urge to rank the races was irresistible. For example, living in the new United States were a million Black Africans, of whom 90 percent were enslaved.[21] Virginia native Thomas Jefferson wrote of them in 1785, before becoming president,[22]

> *Comparing them by their faculties of memory, reason, and imagination, it appears to me, that in memory they are equal to the whites; in reason much inferior, as I think one could scarcely be found capable of tracing and comprehending the investigations of Euclid; and that in imagination they are dull, tasteless, and anomalous.*

I honestly don't know how many Euclid-fluent White people Jefferson knew in the original American colonies, but whatever were his observations and objections to Black people, he had no hesitation continually mating with at least one of them, producing six children.[23]

After Darwin published his seminal 1859 book, *On the Origin of Species*, and especially after his 1871 *The Descent of Man*, we all could have walked forward recognizing that humans are part of one family, holding common genetic ancestry with other apes. We didn't. Instead, many scientists of the day asserted that Black Africans were less evolved than White Europeans. This continued deep into the twentieth century with the 1962 publication of Carleton S. Coon's widely referenced textbook, *The Origin of Races*,[24] where we find: "If Africa was the cradle of mankind, it was only an indifferent kindergarten. Europe and Asia were our principal schools."

The seven hundred pages of text and illustrations offer rampant comparisons between Black Africans and apes. Enough to convince White people to be glad they're born White, leaving them completely open to any rules, laws, and legislation that segregate or subjugate Black people.

When putting forth a scientific hypothesis, you should be your own greatest critic. You don't want colleagues finding holes in your reasoning before you do. It looks bad and it reveals you didn't do your homework. A good way to attack one's own work is to step back and explore whether a completely opposite explanation can be constructed from the same data or from data you may have overlooked. If you succeed at dismantling your own hypothesis, it's time to move on to another research project.

Let's try it.

Reverse the tables: If nineteenth and twentieth-century anthropologists had been Black supremacists instead of White supremacists, what might they have written about White people upon examining them? Your own group must always land at or near the top of your own ranking system, so what observations might support the case that White people are inferior and apelike? And if you're White, how might you feel reading them?

Chimpanzees are humans' closest genetic relatives. We just need to find similarities between chimps and White people, and that would be surefire evidence of their less evolved state.

- Chimps and other apes grow hair all over their bodies. The hairiest people you've ever seen have been White people, with mats of hair across their chests and ascending their backs.[25] Their body hair can even reach upward and out of their shirt collar. Black people do not remotely approximate this level of hairiness.

- Distinct from their face, hands, and feet, part the hair of most chimpanzees—the way they do to each other when checking for lice—and their skin color is white, not any shade of black or brown.[26]

- Chimps tend to have big ears relative to their head size. After decades of ear-watching, I can attest that the biggest ears I've ever seen on humans have been on White people. Have a look yourself, next time you're in a crowded public place. Doubtless, there's strong overlap, but the size of Black people's ears can be as little as half the size of White people's ears. You might now ask about the famously large ears of President Barack Obama, but he is precisely half White—just as much White as Black. So maybe his big ears come from the White half of his heritage.

- For most of the twentieth century, Neanderthals were portrayed as stupid and brutish. Turns out, beginning in the 1990s, genetic research revealed that Europeans are between 1 and 3 percent Neanderthal. Africans, at most, a fraction of 1 percent.[27] That can't be good for Europeans—time to clean up that backward, primitive image. Since then, published references to Neanderthals instead comment on what must have been their creative, artistic, inventive, and articulate ways, crafting sophisticated tools and technologies to shape their world.[28]

- Chimps have extremely thin lips, just like White people. (How did they miss that one?)

Look how easy it is to be racist. Let's continue.

- Chimpanzees invest quality family time preening each other's hair. We've all watched them do this, if not at the zoo, then in television documentaries. Apparently, the lice they find must be tasty because whoever plucks them from the other chimp's head also eats them. Ever hear of a lice outbreak among Black children? Probably not.[29] White children are

30 times more susceptible to lice infestation than are Black children. The parasite simply likes to lay eggs on the hair of chimpanzees and White people more than on the hair of Black people.[30]

What about plain old Black superiority to Whites without reference to apes? How about:

- White people are 25 times more likely than Black people to contract skin cancer.[31] I can count a dozen fully educated White people in my life who argued strenuously that they and I were equally at risk when exposed to sunlight. Superiority is hard to concede when, in your mind's eye, you're the superior one.

- The itchy, scaly skin disease of moderate-to-severe psoriasis is twice as prevalent among White people as among Black people.[32]

- Ever see Black children with so much facial acne that their classmate tormentors called them pizza face? Probably not.

- Ever hear the phrase "Black don't crack"? It refers to the graceful aging of dark skin relative to light skin, with reduced wrinkles and other blemishes. Credit melanin for that.

- Growing up, I had been on several hiking trips during summer camp where most of the White kids got poison ivy rash, and the Black kids didn't. Could just be incomplete data. Maybe somebody should look into that.

- The frailty of old White ladies is legendary and tragic. As they age, one in two of them will ultimately break a bone due to diminished calcium density and the onset of osteoporosis.[33] Old Black bones remain good and strong.[34] If a Black woman breaks her hip, it's because she fell out of a window, not because she slipped on the floor.

- Despite a recent rise in the suicide rate of Black teens,[35] the suicide rate for White people is 2.5 times higher than for Black people.[36]

- Rates are also 2 to 3 times higher for anorexic eating disorders among White women than among Black women.[37]

One more from the chimp vault:

- Chimpanzees love to swing in the trees. Apparently, so do suburban White children. They typically can't wait to build and live in a backyard treehouse. You have not likely seen Black children even contemplate the idea. White people clearly want to return to their fully primitive state.

That racist rant was surely awkward for some to read, in part because the data supporting it are all real. What matters is what motivates you. Do you aim to declare superiority and then act on it in some nefarious way, or are you simply fascinated by the diversity of the human species? In any case, racist White anthropologists conveniently overlooked it all.

How about some authentic racist Black mythology? The skin color of Ethiopians is neither White nor as dark as Black Africans can get. From this fact emerged their origins story, which I paraphrase:[38]

> *God, when baking the lumps of clay that would become humans, removed them from the oven too soon. These are the White people—a failed recipe on His first try. For God's next attempt He left the clay in the oven too long. These became all the dark-skinned Black people. Another failed recipe. In God's third attempt, He pulls the batch from the oven just in time, making the perfect, golden brown skin of Ethiopians.*

We can chuckle, if for no other reason than that they imagine an omniscient, almighty creator who's still learning how to bake, but it's their story. They get to put themselves at the top of their own fantasies, just like everybody else. Meanwhile, next door to the Ethiopians, on the east coast of Africa, are the Somalians. Their skin color is much darker than the Ethiopians', but their facial features are distinctly European. Or rather, the facial features of Europeans are distinctly Somalian. What's a racist person to do about them?

If we give our hypothetical Black racists control over legislation, they may just do what the White racists did against Black people—create laws that prevent White people from being educated and then justify their subjugation and enslavement for being stupid and subhuman.

Africa is indeed the "cradle of mankind." A few hundred thousand years ago, early humans wandered north and then west and east, populating Europe, Asia, and ultimately the Americas. Our peripatetic ancestors carried the base African genome all around the world. Those journeys took less time than you might think. Let's do the math. If you walked 25,000 miles (Earth's circumference) at 2 miles per hour—a leisurely pace—and you did it for eight hours per day, you would circumnavigate the globe in 4.3 years. Of course there's no actual path or road that goes around Earth, and there's desert and bodies of water and mountains and other stuff in the way. Nonetheless, before agriculture, generations upon generations of Africans, one after another, had plenty of time to reach every contiguous nook of Earth's surface—driven by the search for food or by a sense of wonder about what lay beyond their horizons.

Today, one out of six people in the world lives in Africa. A continent that's five times larger than Europe and home to fifty-four countries, more than one-fourth of the world total. Upon seeing its residents, if all you notice is dark skin, then you're missing the most important features of who lives there. As the source of human origins, the African continent manifests the greatest genetic diversity of any place on Earth, provided you look past the skin color. With that diversity comes taxonomic variance. Where in the world would you expect to find some of the shortest and the tallest people? Africa. The Mbutsi Pygmies from the Democratic Republic of the Congo average just over four feet tall.[39] And among the Watussi of Rwanda and Burundi, both adjacent countries to the Congo in Central Africa, the average man stands six feet tall.[40]

Short and tall genetics exists elsewhere in the world, but they do not occupy the same geographic area. The average height of those native to the Netherlands is seven to eight inches taller than that of Indonesians,[41] but they're separated by nearly a third of Earth's circumference.

Where might you find some of the world's slowest and fastest runners? Athletic contests generally do not seek out slowpokes. Among fast runners, Africa and its emigrant descendants have dominated the international stage of track and field in both sprints and long distance for most of the past century.

How about very stupid and very smart people? Likely find them both in Africa too. Let's focus on the smartest for now. Time to be reminded that Egypt and its towering civilization of engineering, architecture, and agriculture predated Europe's by thousands of years. Last anybody checked, Egypt is in Africa. The civilization was so advanced that denial among White people runs deep,[42] including the storyline of *Stargate*, a 1994 sci-fi film in which the Great Pyramids were conceived and designed not by Africans, but by godlike humanoid aliens who subjugated the human Egyptians. Space entrepreneur Elon Musk (an African himself)[43] even Tweeted as such on July 31, 2020: "Aliens built the pyramids obv." From the fifteenth century onward, European explorers, colonists, slave traders, and traveling anthropologists never went looking for people smarter than themselves in Africa, although the continent was likely to have no shortage of them. You just have to look. Neil Turok, a White South African physics colleague of mine, founded in 2003 the African Institute for Mathematical Sciences (AIMS).[44] It offers inspired, advanced-level teaching and support for graduate degrees in math, engineering, and physics. Though based in South Africa, the organization serves all countries on the continent. Beginning in 2008, AIMS's mission expanded to include a targeted search to "unlock and nurture scientific talent across Africa, so that within our lifetimes we are celebrating an African Einstein."[45] They do so by pursuing stories from schoolteachers, professors, and elders across the land, about any students in their communities whom they find to be uncommonly clever and who may benefit from this opportunity.

Nobody denies the smarts of people who are great at the game of chess. With such a low financial barrier to entry, chess is played and contested worldwide, no matter the wealth of the country. Time to note that the average rating of the top-ten chess players in Zambia, the heart of Africa, is higher than those of Luxembourg, Japan, United Arab Emirates, and South Korea.[46] Let's peek at each of these countries' 2020 GDP per capita: Luxembourg $116,000, Japan $40,000, the United Arab Emirates $36,000, South Korea $32,000, Zambia $1,000.[47]

On May 1, 2021, a talented chess player reached the title of National Master for having achieved a US Chess Federation rating above 2200, landing among the top 4 percent of 350,000 total rated players in the world.[48] A rating achieved that was five hundred points higher than that of his chess coach, just a few years after learning how to play the game. That prodigy is a ten-year-old boy named Tanitoluwa Adewumi, the son of Nigerian refugees to the US in 2017. His family spent a brief time living in homeless shelters in New York City before his parents established stable employment and permanent residence. I played a brief chess game against the little fellow in March

2021 on Grandmaster Maurice Ashley's Twitch platform, a livestreaming social media interface. The game was indeed brief.

Speaking of Nigerians, immigrants to the US enjoy an 8 percent higher household income[49] than the national median. And ethnic Nigerian children in the United Kingdom, especially those from the Igbo tribe, consistently attain higher test scores on average than their White UK counterparts.[50]

Occasions to pause and wonder what depths of intellectual capital in math, science, engineering, or any field lay hidden deep within the African continent, or anywhere else on Earth—lost for now or lost forever—for want of opportunity to flourish.

Consider further what happens within the human ancestral tree. Overhearing someone in New York City say he's Italian, I asked where he was born. "Brooklyn." Where were his parents born? "Brooklyn." Where were their parents born? "Italy—so I'm Italian." In conversation with another person who declared she was Swedish, I asked where she was born. She replied, "New York City." I kept going. Where were your parents born? "Minnesota." Where were their parents born? "Minnesota." And their parents? "Sweden—like I said, I'm Swedish." You can see what's happening here. In these two cases, they both could have said, "I'm American," but instead journeyed up the family tree to cherry-pick the place that pleases them most. Their stopping point was arbitrary, so I invited them to keep ascending their family trees until they reached Africa. Because in the end—or rather in the beginning—we're all African.

Human ancestry is stupefyingly convergent. Of the 8 billion people in the world, everyone has a single pair of biological parents. Imagine if couples bore only one child. Then that child's parents' generation would number 16 billion people. If their biological parents had also borne just one child, then their generation would number 32 billion people. If their biological parents also had only one child, then their generation would contain 64 billion people. We now span four generations, back to the year 1900. We don't expect more than three generations to be living at the same time. So in 1900, that gives 64 + 32 + 16 = 112 billion people. Earth's actual population in the year 1900 was fewer than 2 billion.

What's happening here? Moving back through time, 112 billion people have somehow converged to 2 billion. Going back further, in the year 1800, the population was 1 billion. In 1600, only 500 million. At the time of the Egyptian pyramids, no more than 20 million people existed in the world.[51] That equals today's population of the New York metropolitan area.[52] The only way to reconcile these numbers is to rapidly funnel highly "unrelated" people down into fewer and fewer families. Not to mention that in any generation many people are siblings, and upwards of 20 percent of all people have no descendants at all.[53] This exercise reveals that any two people on Earth share a common ancestor who sits at a node of a rapidly converging family tree. In genealogy, this phenomenon is called pedigree collapse. That's also how hundreds of millions of White people can proudly and legitimately claim to be descendants of Charlemagne (c. AD 800). If better state records were kept of peasant babies and orphans, we'd surely find common ancestors there too, otherwise overlooked in the art of cherry-picking your family tree.

When I imagine what I'm capable of achieving, I don't reference the professions of ancestors reported to me in a genealogy kit. Instead, I look to all humans who have ever lived. We are one family. We are one race. The human race. Although I rather think we're all just next of kin.

Whether or not day-to-day feels this way, civilization has made great social advances over the decades and centuries. Progressive changes in laws, legislation, and attitudes around the world have, in some areas, brought the diversity of race and gender to social levels that approximate Martin Luther King Jr.'s famous line from his 1963 "I Have a Dream" speech:[54]

> *I have a dream that my four little children will one day live in a nation where they will not be judged by the color of their skin but by the content of their character.*

One last thought experiment bears this truth. Imagine you're offered the opportunity to enter a time machine and go back to any previous moment in human history. Where and when would you choose? The White, cisgender, heterosexual male has his pick of the timeline—this traveler to the past will be welcomed everywhere and everywhen. If that's not you, then you'd better think hard about the time and place you'd like to arrive. Are you female, a person of color, disabled, queer, or any combination of these? When was it better for you? A thousand years ago? Five hundred? A hundred? Fifty years ago? Ten? Five? For me, I'm good with the present. I'd rather not be declined a taxi ride, overlooked for job opportunities, denied a bank loan, or redlined from my choice of housing. With childhood ambitions of being a scientist, I don't want to be someone's servant, nor do I fancy being purchased and owned by another human being who thinks I'm not entirely human. Come to think of it, I'd rather visit the future, as I presume the preternaturally progressive Unitarian minister Theodore Parker would too, as he wrote in 1853:[55]

> *I do not pretend to understand the moral universe; the arc is a long one, my eye reaches but little ways; I cannot calculate the curve and complete the figure by the experience of sight; I can divine it by conscience. And from what I see I am sure it bends towards justice.*

Do we recognize, highlight, and embrace diversity? Or do we aspire to not notice it at all? Imagine if race, gender expression, and ethnicity were as irrelevant to our judgments of people as whether they wear glasses, what brand of toothpaste they use, or whether they prefer waffles over pancakes.

My sentiments align with the visiting alien on this one. Consider extraterrestrials so different from us in every way that to them all humans are indistinguishable from one another, no matter how much we distinguish ourselves. All they see of us are four limbs, a torso, and a head. Sounds insensitive of them, but we're no better. For most animals on Earth, we have no idea at a distance and probably not even close up what their gender is or whether there's some subtle difference between the coloration or plumage on one member of a species relative to another. We silently think that way about urban pigeons, and especially suburban goldfish. Parents surreptitiously swap live ones for the dead ones they've just killed. This typically happens while their kids are away at camp as they attempt to conceal that they've overfed (or neverfed) Goldie and Bubbles. For most of us, seen one goldfish, seen 'em all.

Our visiting alien sees us segregate, stratify, and subjugate others among us based on features they hardly notice. Bearing witness to our divided ways—in response to all that should be irrelevant to the content of our character—our space alien would surely phone home and report further evidence that there's no sign of intelligent life on Earth.

NINE

LAW & ORDER

The foundation of civilization, whether we like it or not

If you kill an alien, is it murder? Does the alien have to be more intelligent than you for its death to be classified as a homicide? Who owns the Moon, the mineral rights to asteroids, or the water rights to comets? Which nation's laws govern one patch of a planet's surface relative to another? Space law remains a bit of the Wild West relative to established doctrines on Earth. If we want our conduct in space to be no worse than our conduct on Earth, it will require enforceable laws, perhaps jump-starting arcane justice systems that haunt the past. For all these reasons, space law is itself a frontier of legal philosophy.

Legal systems at their best are prerequisites for anything we call civilization, as they protect us from the destabilizing basal urges of our own primal instincts. Ask yourself, how would people behave if no threat of legal action existed in this world? Look at how many people transgress established laws even with a legal system in place. Without it, there's not much hope for civilization.

If Aristotle's edict matters: "Law is reason free from passion," and if the allegorical Lady Justice, blindfolded, wielding a sword in one hand and balance scales in the other, truly symbolized how it works, then your jury every time would contain data and information experts who, in search of the truth, each had enhanced immunity to the passion of lawyers, the emotion of witnesses, and the force of public sentiment. If legal trials are about speaking the truth, the Latin root of which is "verdict," and judgments are the meting out of punishment or rewards, then why are some lawyers higher paid than others? Are they better at finding the truth? Or are their methods and tactics simply better at passionately persuading a jury of whatever they want, regardless of what is true?

In the court of law, if truth and objectivity are neither sought nor desired, then we must admit (confess?) to ourselves that at least some parts of the justice system are the opposite of Aristotle's edict, and are instead all about feelings and emotions. A quest to turn passion into compassion. The consequence? Some fraction of the time, a verdict arrives that does not represent the truth as much as it represents what high-paid attorneys needed to be true.

Consider the evolution of trials, simplified here for the sake of storytelling. We begin with a person in charge declaring with or without evidence whether you are guilty as accused. They use their own judgment—what feels right to them—and declare a verdict. No adjustments for bias, bad mood, or misinformation.

Surely, we can improve on that.

In religious cultures, deities typically see and know all things. Santa Claus's lyric is a modern variant: "He knows when you've been bad or good, so be good for goodness' sake." So why not let God decide the verdict. In trials by ordeal, you might be forced to engage in one-on-one combat, get dunked in water, walk through fire, have boiling oil poured on your chest, or drink poison. If you survive these with minimal injury, you must be innocent because God protected you. Examples of such trials span cultures and trace back as early as the Code of Hammurabi in ancient Babylon, 1750 BC. For example, number 2, out of 283 laws, describes trial by water:[1]

> *If any one bring an accusation against a man, and the accused go to the river and leap into the river, if he sink in the river his accuser shall take possession of his house. But if the river prove that the accused is not guilty, and he escape unhurt, then he who had brought the accusation shall be put to death, while he who leaped into the river shall take possession of the house that had belonged to his accuser.*

Nothing about that law looks good. In a variation on this theme, let's toss your dead body in the ocean. If you float faceup, that's evidence God took your soul, and your innocence is rewarded in Heaven. Facedown, you're guilty as charged, and heaven doesn't want you.

You can't make this stuff up. In chapter 48 of Antonio Pigafetta's eyewitness account of Magellan's round-the-world voyage,[2] he recounts a particularly rough two months at sea without fresh food or water. As is custom, dead crew members are discarded overboard into the ocean. Pigafetta describes a scene reminiscent of trial by water:

> *We sailed northwest for two months continually, without taking any refreshment or repose. And in that short space of time, twenty-one of our men died. And when we cast the Christians into the sea, they sank with face upward toward heaven. And the Indians always with face downward.*

Either this really happened as described, or Antonio succumbed to a bad case of confirmation bias. In either outcome, if you are going to use the faceup-while-dead-in-the-water evidence to establish guilt in the court of law, then everybody has to be Christian for it to work as intended. What if they are not Christian? Then the accused simply end up dead, whether they're guilty or innocent, with no access to Heaven.

Why not let evidence decide? The accusers bring forth evidence to convince the one who judges. Once again, suppose the one who judges doesn't like you. Suppose you are innocent but nobody argues on your behalf. Evidence only works if the people examining the evidence know how to do it and care about what is true.

In all fairness, Sr. Pigafetta was a gentleman of the 1500s—a pre-scientific era. As already noted, hypothesis testing as practiced today did not become a routine thing in science until the 1600s. Before then, natural philosophers—what today we call scientists—were perfectly content declaring things that seemed true, ignorant of a later time when experiments and observations would supplant suppositions. Not humans, but nature—the universe itself—would become the ultimate judge, jury, and executioner in science. The nineteenth-century naturalist Thomas Henry Huxley expressed the same thing more bluntly:[3]

> *The great tragedy of Science—the slaying of a beautiful hypothesis by an ugly fact.*

Why not have the public decide—the wisdom of the crowds? The verdict would no longer be susceptible to the mood of the sitting judge. But wait, crowds can be shockingly irrational. They can morph into mobs where the sum of their collective brain power dilutes as the size of the mob grows. Is there really any other occasion in life when you would chant angrily while wielding a pitchfork and torch? Crowd justice is the precise recipe for lynchings. Today, crowds cancel people on the internet; welcome to the court of public opinion and not the court of law. A tally of up and down votes, where people get to weigh in on what they want to be true, or what ought to be true, and not on what is actually true. To find out what is actually true requires careful and thorough investigations and not the simple act of reading media accounts and forming your opinion based on them.

Here's a better idea. Let's collect a set of ordinary people, not too many, not too few. Expose them to the evidence for and against the defendant and grant them the power to decide guilt or innocence. This avoids the whims of a despot, the bias of a single judge, the gore of trial by ordeal, and the mindlessness of a mob. Suppose the jury comprises people who hate you for any reason. Or suppose, because of their station in life relative to yours, they have no understanding of your circumstances. So maybe they don't hate you—they just don't care about you. That's bad too.

Here's an even better idea. Let's compose a jury, not of just anybody, but of your peers. This gives you the best chance of getting a fair trial, minimizing or eliminating potential bias against you, although there could be bias in your favor if indeed you are guilty. That's an acceptable risk, as the famed Blackstone ratio declares:[4]

> *It is better that ten guilty persons escape than that one innocent suffer*

which was first expressed in the 1760s by the English jurist Sir William Blackstone.

We now arrive at what is fundamentally the modern system of justice in the Western world: trial by a jury of your peers. Holding aside more than a century of lynchings in the US that began in the 1830s, the Sixth Amendment of the US Constitution, fully ratified by 1791, encapsulates this urge to protect the accused:

> *In all criminal prosecutions, the accused shall enjoy the right to a speedy and public trial, by an impartial jury . . . to be informed of the nature and cause of the accusation; to be confronted with the witnesses against him; to have compulsory process for obtaining witnesses in his favor, and to have the assistance of counsel for his defense.*

These ideas go way back, even before King John's Magna Carta of 1215, which contains the declaration that no free man may suffer punishment without the "lawful judgment of his peers."[5] Today, our system of justice continues to orbit these principles. In theory, everyone should get a fair trial, but in practice a high-profile attorney can sway a jury to feel one way or another, on a level that influences their interpretation of the data, and thus, in a choice moment, sow bias in the courtroom—bias that might not have otherwise been present when the trial began.

In fact, we breed future lawyers to wield such powers of persuasion. Fertile feeders to the legal profession include debating clubs in high schools and colleges across the country. Was never much interested, myself, even though my high school fielded a particularly competitive and successful debate team. Their hallway cabinets were chock-full of trophies. Not so much for our sports teams. Regardless, I was a full-on adult before I learned what actually happens at these tournaments. The debaters are given the topics that will be covered, except they don't know what side of the debate they will be assigned to argue until the day of the contest. The winner is the person or the team that makes their case most convincingly to the judges. The goal is not to find out what is objectively true about anything. Instead, the entire system trains you to argue—either side of any subject. The rules also presuppose that all topics of debate have only two sides to them. Not three. Not five. Not a continuum. Learning of this, as a budding scientist, I could think of nothing more disruptive to the search for truth than this breeding ground of arguers.

Maybe I'm overreacting here. I attended the Bronx High School of Science in NYC, a source of eight Nobel Prizes among its graduates—seven in physics, one in chemistry.

The culture of debate couldn't have badly hurt our culture for science. It nonetheless leaves you wondering. Our political representatives, from the beginning, have been heavily drawn from the legal profession. They are collectively called "lawmakers." Do the persistent impasses in congressional deliberations owe their origins to the art of arguing rather than the science of truth-finding?

Still, the attempt to determine guilt or innocence has progressively evolved over the ages. That's a good thing, we can all agree. Is there room for further improvement? Does the system want further improvement? Seems like the answer to that question is no.

During my first-ever call to jury duty, I was also on the teaching faculty of Princeton University, where I designed an undergraduate seminar on what science is and how and why it works. During the question-and-answer session that attorneys conduct with prospective jurors—unhelpfully called voir dire—I was asked what magazines I read, what TV shows I watch, where I get my news. I was then asked what I did for a living. "I'm a scientist." Knowing from the written questionnaire that I teach at Princeton, they asked what I teach. "I teach a class on the evaluation of evidence and the relative unreliability of eyewitness testimony." I did not survive that round and was on my way home within the hour.

When scientists overhear the dramatized courtroom call of "I need a witness!" we think to ourselves, "For what?" Psychologists fully understand this disrespect of eyewitness testimony.[6] Two sane people can observe the same events or phenomena and report them differently with equal sincerity and confidence in their own account. The more extraordinary or shocking the event—like witnessing a violent crime or greeting a space alien—the less likely the various accounts of the experience will match. That's why the methods and tools of science were invented in the first place—to remove human sensory frailties from the acquisition of data. Eyewitness testimony may sit high in the court of law, but it sits low in the court of science. If you show up at a conference and the best evidence for your research is that you saw it happen, we will show you the exit door.

In my second call to jury duty, the judge read the basics of the case; the guy being charged was right there in the room. Drug possession. Cocaine. Manhattan. The defendant was charged with possession of 1,700 milligrams of cocaine, sold to an undercover narc. We arrived at the juror interrogation part, and I was asked whether I knew any lawyers. At the time, I had no lawyer friends. When we got to the end, the judge intervened and asked, "Are there any questions you'd like to ask the court about this process?" I raised my hand and said, "Yes, Your Honor, why did you say that the defendant was in possession of 1,700 milligrams of cocaine? The thousand cancels with the 'milli' to get 1.7 grams, which is less than the weight of a dime." When I said this, everyone in the courtroom looked my way and nodded their head. I continued, ". . . so it looks like you are making the quantity of drugs sound like more than it actually was."

I was back out on the street once again, within the hour.

I would later wonder whether my question contaminated other prospective jurors in the room. Regardless, do any of us ever say, "I'll see you in sixty-billion nanoseconds"? No. We say instead, "I'll see you in a minute."

On my third call to jury duty, the case was a robbery. A literal "he said, she said" case. A man was accused of robbing a woman of her groceries and her purse. When the police found the assailant shortly afterward, he was positively identified by the victim, but he was not in possession of what was claimed to have been stolen. So this complicated matters greatly. In this round of jury selection, I had reached the final fifteen—the closest I had ever come to serving on a twelve-person jury. Then the judge asked us, one by one, would we have a problem reaching a verdict on the kind of evidence presented. I replied, "Based on all I know of the unreliability of eyewitness accounts, if the only evidence is eyewitness testimony, with no material evidence to support it, I cannot vote to convict." The judge then brought my objection to the rest of the group, saying, "Does anyone else feel the way he does, that you need more than one eyewitness before you can arrive at a verdict?" Immediately, a prospective juror seated in front of me declared, "That's NOT what he said!" In that moment, I (successfully) resisted with all my might from saying, "Your Honor, you were eyewitness to my words just a few billion nanoseconds ago, and you got them wrong." Even so, I was, once again, back on the street within the hour.

If an attorney's power to sway a jury with passionate arguments, regardless of the data, is precisely the legal system you want, then you will never, ever want me or any of my fellow scientists on a jury. You will never want any expert in data analysis, in statistics, or in probability. You likely won't want engineers either. If the current system of justice happens to be what we've got, en route to a better system, then there's indeed room for improvement here. Maybe we shouldn't be content with "The system is flawed, but it's the best we've got," even though the exploitation of these flaws makes excellent courtroom storytelling for stage, television, and film. For example, in the 1954 drama *12 Angry Men*, written for TV and later for cinema by Reginald Rose, a lone juror, who does not jump to conclusions from the evidence presented, slowly and rationally uncovers layers of bias that include ageism, ableism, and racism in each of his eleven fellow jurors. It's a murder trial. So one should take a little extra time arriving at a verdict. By the end of the story, which unfolds entirely in the deliberation room, each of the eleven jurors changes his vote to not guilty. Had the jurors been rational, analytic people with minimal to no bias, the story would have never been conceived. Or it would have lasted ten minutes.

Another consequence of the modern legal system is the mere existence of the Innocence Project, founded in 1992. Their mission statement says all you need to know:[7]

> *The Innocence Project's mission is nothing less than to free the staggering numbers of innocent people who remain incarcerated, and to bring substantive reform to the system responsible for their unjust imprisonment.*

Since 1973, more than 186 people who were sentenced to death in the US have been exonerated. A total that would have otherwise been added to the 1,543 prisoners on death row who were executed over that time.[8] And since 1989, DNA evidence alone has freed 375 wrongfully accused prisoners across thirty-seven states who have served a total of 5,284 years behind bars.[9] Using the 10:1 arithmetic of Blackstone's ratio, and applying it to time served, we can ask whether 52,840 years of unserved prison time by guilty people who went free justifies the 5,284 years of time served by the innocent?

With DNA and the burgeoning field of forensics, it seems

like science has come to the rescue. Problem is, the presentation of evidence is still embedded in a system that retains the power to manipulate jurors' biases, independent of what is true. This spawned a study by the National Academy of Sciences on the rampant abuses of science in the courtroom. A 348-page 2009 report, titled *Strengthening Forensic Science in the United States: A Path Forward*, contains the following paragraph in its summary:[10]

> *In some cases, substantive information and testimony based on faulty forensic science analyses may have contributed to wrongful convictions of innocent people. This fact has demonstrated the potential danger of giving undue weight to evidence and testimony derived from imperfect testing and analysis. Moreover, imprecise or exaggerated expert testimony has sometimes contributed to the admission of erroneous or misleading evidence.*

As a capstone to all this, the best-selling author Alice Sebold wrote a 1999 memoir titled *Lucky*, in which she identified the Black man who she said raped her in 1981 at age eighteen. He was later arrested, convicted, and imprisoned for sixteen years—until the end of November 2021, when he was exonerated after a re-review of the evidence against him. A week later, Sebold's book was pulled by Scribner, her publisher. She has since apologized to her victim for misidentifying him.[11]

The Innocence Project reports that 69 percent of the exonerated cases involved witness misidentification, including in-person lineups, in-court appearances, mug shots, police artist sketches, and voice misidentification.

Independent of all demographics that might be cited for incarceration rates, such as race, age, religion, poverty, employment, broken home, etc, fully 93 percent of all prisoners in the US,[12] and worldwide,[13] have a particular trait in common. Extensive studies of their genetic profile reveal they carry a Y chromosome.[14] Nearly all humans who have started wars also carry this socially regressive trait. Yes, it's a man problem. If we could somehow repair the flaws in their genetic code, the world would be a much safer place for us all. We might instead blame testosterone, but many of our greatest emblems of nonviolence, including Jesus, Mahatma Gandhi, and Martin Luther King Jr., were men. Further, most of the world's men will not commit a felony their entire lives, leaving us with yet another unsolved mystery of the universe.

Maybe what Earth needs is a rational virtual country—a solution to the irrational conduct that currently drives crime and punishment and world politics. That's what Silicon Valley entrepreneur and marketing executive Taylor Milsal proposed at a cocktail reception of the 2016 Starmus science festival[15] in Spain's Canary Islands. The concept of Rational Land, as she called it, caught everyone's attention. A bunch of us at this festival, including noted scientist-educators Brian Cox (particle physicist), Jill Tarter (SETI researcher), Richard Dawkins (evolutionary biologist), Jim Al-Khalili (theoretical physicist), and Carolyn Porco (planetary scientist), discussed potential charter municipalities based on which places might jump at the opportunity. Member states would embrace rational thinking in their conduct and policies. Obvious candidates included the cities London, Paris, and New York. The countries Switzerland and Denmark. The US states of Massachusetts, Minnesota, and California. The conversation spread rapidly through the cocktail party, each person adding a bit here and there to what Taylor started.

For me, the name Rational Land didn't roll off the tongue, so I suggested Rationalia. I also felt that annexing entire municipal populations would overlook wildly irrational factions that may be operating within them, as well as omit highly rational factions that may operate outside of them. And so, after extensive discussion, primarily with Brian and Jim, we instead embraced the individual's choice of virtual citizenship, which happens to be ideal for social media, and which led to a simple Tweet that I posted during the conference:

Neil deGrasse Tyson
@neiltyson

Earth needs a virtual country: #Rationalia, with a one-line Constitution: All policy shall be based on the weight of evidence

10:12 AM · Jun 29, 2016 · TweetDeck

There would be no citizenship tests. No immigration rules. No pledges of loyalty to take up arms against enemies. Just resonance with that one-line constitution.

Shortly after this posted to my followers, some respondents went apoplectic. I was stunned by how many organizations and media outlets hated the idea, being sure that a country founded on evidence and rational thinking would not work. Some headlines:[16]

U.S. News & World Report
THE RATIONALIA FALLACY

Slate
A NATION RULED BY SCIENCE IS A TERRIBLE IDEA

The Federalist
NEIL DEGRASSE TYSON'S "RATIONALIA" WOULD BE A TERRIBLE COUNTRY

Arts Journal
SORRY, NEIL DEGRASSE TYSON, BASING A COUNTRY'S GOVERNANCE SOLELY ON THE "WEIGHT OF EVIDENCE" COULD NOT WORK

Wow. Two of those four media outlets used the word "terrible" in their titles, making sure upfront that their opinion becomes your opinion before you even read their article. The *Arts Journal* headline, beginning with the word "Sorry," is how grown-ups speak to children who have just offered a crazy idea, but you, being the adult in the room, must politely tell them it won't work. With that much power of opinion aligned and mounted against not only me, but the concept, and by association my fellow scientist-educators and Taylor Milsal, we all must be completely clueless, if not insane.

Or maybe railing against people you disagree with takes less effort than exploring why they think differently from you. None of those media outlets contacted me in advance for comments to include in their articles. They were not interested in dialogue. My social media following happens to be large enough that I could post a reply addressing all their concerns and have it reach a larger audience than their combined circulations. Just sayin'.

I recoiled at the persistent tone that scientists should not shape geo-policy. The most virulent objection was the question of where such a country would get its morals and how other ethical issues might be established or resolved.

The last time I reviewed the US Bill of Rights, there was no discussion of morals there either. Nowhere does it say "Thou Shalt Not Murder." Yet there's an entire amendment—number 3—that prevents the military from bunking in your home without your permission.

And if court verdicts were fully evidence-based, then, inspired by the Scottish justice system, we're prompted to redefine "not guilty" and add a third verdict of "innocent":

Guilty: Evidence shows you committed the crimes as accused.

Not Guilty: We think you are guilty but can't prove either your guilt or innocence.

Innocent: Evidence shows you did not commit the crimes as accused.

Also, consider that morals evolve across time and culture, typically by rational analysis of the effects and consequences of previously held morals, in the light of emergent knowledge, wisdom, and insight. The Bible, for example, often held up as a source of morality, is not a fertile place to find antislavery commentary, nor discussions of gender equality.

The Rationalia Tweet specifically references policy, which

can more broadly set frameworks for thinking about laws. Examples of policy would be a government's choice to invest in research and development, and if so, by how much. Or whether a government should help the poor, and if so, in what ways. Or how much a municipality should support equal access to education. Or whether tariffs should be levied on goods and services from one country versus another. Or what tax rate should be established, and on what kinds of income. Or whether "carbon credits" should be implemented to manage and ultimately discourage the use of fossil fuels. Often these policies stall between political factions arguing loudly that they are right and their opponent is wrong. Which calls to mind the mostly true adage, "If an argument lasts longer than five minutes, then both sides are wrong."

Further, the Rationalia Constitution stipulates that a body of convincing evidence needs to exist in support of an idea before any policy can be established that is based on it. Any absence of data can itself be a source of bias. In such a country, data gathering, careful observations, and experimentation would be happening all the time, influencing practically every aspect of our modern lives. As a result, Rationalia would lead the world in discovery because discovery would be built into the DNA of how the government operates and how its citizens think. The absence of relevant data would also be a well-known source of bias.

In Rationalia, the sciences that study human behavior (psychology, sociology, neuroscience, anthropology, economics, etc.) would be heavily funded since much of our understanding of how we interact with one another derives from research within subfields of these disciplines. Because their objects of study are humans, these fields are particularly susceptible to social and cultural bias. So for them especially, the verifiability of evidence will be of highest concern and priority.

In Rationalia, if you want to fund art in schools, you simply propose a reason why. Does it increase creativity in the citizenry? Does creativity contribute to culture and to society at large? Will creativity matter in your life, no matter your chosen profession? These are testable questions. They just require verifiable research to establish answers. The debate ends quickly in the face of evidence, and we move on to other questions.

In Rationalia, since weight of evidence is built into the Constitution, everyone would be trained from an early age in how to obtain and analyze evidence and how to draw conclusions from the data.

In Rationalia, you would have complete freedom to be irrational. You're just not free to base policy on your ideas if the weight of evidence does not support it. For this reason, Rationalia might just be the freest country in the world.

In Rationalia, citizens would pity newscasters for presenting their opinions as facts. Everyone would have a heightened capacity to smell bullshit wherever and whenever it arose.

In Rationalia, for example, if you want to introduce capital punishment, you'd need to propose a reason for it. If the reason is to deter murder, then an entire research machine would be put into place (if it did not already exist) to see whether, in fact, capital punishment deters murder. If it does not, then your proposed policy fails, and we move on to other proposals. If capital punishment does, indeed, deter murder, then you must follow with the question: If the state is granted the power to take the life of its own citizens, and does not also wield magical powers to bring them back to life, then what happens if you later discover that the person you executed was innocent of the crime?

In Rationalia, a diverse, pluralistic land, you are free to practice religion. You would just have a hard time basing policy on it. Policies, by most intended meanings of the word, are rules that apply to everyone, but most religions have rules that apply only to their followers.

In Rationalia, research in psychology and neuroscience would establish what level risks we are all willing to take, and how much freedom we might need to forfeit, in exchange for comfort, health, wealth, and security.

In Rationalia, you could create an Office of Morality, where moral codes are proposed and debated. What moral codes would the citizens of Rationalia embrace? That is, itself, a research project. Countries don't always get it right. Neither will Rationalia. Is the enslavement of people with dark skin an okay thing? The US Constitution thought so for 76 years. Should women vote? The US Constitution said no for 131 years.

And if we learn later that Rationalia's constitution needs additional amendments, then you can be sure there will be evidence in support of it. In such a world, people will surely still argue, but they won't likely go to war over their differences of opinion. Courts would be bastions of rational discourse, rendering courtroom dramas to be the most boring genre of television ever conceived. These may be the foundations of eternal justice and everlasting peace. Or maybe it won't be perfect for reasons yet to be gleaned, but it will be the best we have so far.

In the end, it comes down to passing and obeying laws

that we all respect. Laws that require objectively verifiable evidence to prosecute offenders. Laws that we all think are just. Laws that we hope will advance the interests of civilization. And laws that promote harmony in the population rather than discord. Also, if Miss Demeanor commits a mild infraction, you want to understand why so that the causes can be assessed and rectified, circumventing future transgressions. This does not always involve punishment. For such goals to succeed requires a system of laws based on objective truths that apply to everyone, rather than on political or personal truths that apply only to some.

Let's all hope for people in the not-too-distant future that today's system of justice will look to them what trial by death-dunking looks like to us.

TEN

BODY & MIND

Human physiology may be overrated

Some of my best friends are made of chemicals. Actually, all of my best friends are made of chemicals. We badly want the human body, and perhaps life in general, to be more than the sum of its electro-biochemical reactions. Whether or not you are religious, references to a human "soul" or a "spirit energy" or a "life force" represent common examples of what feeds these desires. Regardless, the chemistry part of us remains in full swing. The annual *Physician's Desk Reference* (*PDR*)[1] is a massive compendium of all prescription drugs available—more than a thousand—complete with listings of pharmaceutical producers, color images of the medications, recommended doses, side effects, contraindications, chemical formulas, and any other pertinent information for the medical doctor. Add that to the seemingly uncountable number of over-the-counter nonprescription drugs, dietary supplements, and herbal treatments, and you've got an entire sector of the world's economy that creates and supplies chemicals for your well-being. Herbal treatments, whether ancient or modern and whether they work or not, still constitute an infusion of chemicals into your body. They just weren't made in a laboratory. To live an affliction-free life, we must confess to nature that we're a sack of chemicals occasionally (or frequently) in need of help from other chemicals to live our fullest life. Given how often illness befalls us, from childhood through old age, and given how often our body parts malfunction, perhaps we should instead be amazed that the human body works at all.

If so, then how amazed should we be?

My seventh-grade science teacher was a big fan of the human body. He was especially fond of the heart, which can pump for eighty years and longer without shutting down: "No machine we've ever built has lasted that long without repair." He also praised our hands and feet, describing them as pinnacles of evolutionary design, with bones and muscles and tendons and ligaments, all in the right places. The *Vitruvian Man*, famously illustrated by Leonardo da Vinci in 1490, helped establish this ideal. It shows a human form, with arms outstretched, embedded in a perfect circle, capturing perfect geometric proportions within it. The exact center of the circle? The human navel. These seemed like strong arguments at the time—I was twelve—but I would later learn that the location of human belly buttons enjoys huge variance, and you're dead within a week if you consume no water. Catastrophic organ failure leads to a stoppage of your beating heart.[2] So no, the heart does indeed require constant maintenance. We just don't think about it that way.

As for the marvel of our two feet, each containing twenty-eight bones and attendant ligaments and tendons,[3] competitive runners who have no feet use curved blades attached to their legs. You've surely seen them in the Paralympics. They look nothing like the human foot, yet are better designed and are more energy efficient for walking and running. For these reasons and countless others, there is little incentive to invent intelligent robots that look exactly like us, given the flaws and shortcomings of the human form.

Though robots may harbor unwelcome computer viruses, we harbor welcomed microbes. Lots of them. More bacteria live and work in every centimeter of our lower colon than the sum of all humans who have ever lived. To them, we're nothing more than a warm anaerobic vessel of fecal matter. Who's in charge? We are, mostly. Unless you disturb the microbes—throwing them out of equilibrium. Then they're in charge, ensuring you know at all times your distance to the nearest toilet. Tally them all, living synergistically and symbiotically in our gut and on our skin, and you get more living organisms than cells of our own bodies.[4] The number may be as high as 100,000,000,000,000 (one hundred trillion) microbes. Some of them may even influence what foods we crave, such as chocolate, as they break down larger molecules into smaller ones that more easily pass into your bloodstream.[5] You think your cravings are your own. Instead, the chocoholic bacteria in your gut are what's summoning the bonbons.

What about our senses? The human body, complete with its complex biological and electrochemical pathways, is all we've got to decode the environment. The five traditional senses—sight, hearing, touch, taste, and smell—are rightly cherished for their capacity to detect external stimuli. So valued are these probes of our world that if you're missing any of them, you're considered handicapped.

Ranked by distance, sight comes in first place. The farthest thing visible to the human eye is a twin of our own Milky Way, the Andromeda Galaxy, which sits two million light-years away, far beyond the stars of the night sky. Next is our sense of hearing. If a sound started loud, such as a thunderclap, you could hear it all the way from your horizon, several miles away. As for our sense of smell, you can usually tell from anywhere in a home if somebody just burnt the dinner, although smoke detectors have admirably usurped that role. Lastly, to taste or touch things requires that they come in direct contact with your body.

Science itself did not reach experimental maturity until engineers developed tools to hone, extend, and even replace each of our five senses, themselves highly susceptible to our accompanying mental states. Not only that, we've discovered senses that lie far beyond human physiology. Indeed, our five biological senses pale when compared with the dozens of "senses" that science now wields, each offering extraordinary access to the operations of nature. We fully detect otherwise invisible electromagnetic fields, including radio waves, microwaves, infrared, ultraviolet, x-rays, and gamma rays. We measure gravitational anomalies, the polarization of light, the spectral decomposition of light, chemical concentrations of parts per billion, barometric pressure, and atmospheric composition. In the hospital, we've got MRIs. The abbreviation stands for magnetic resonance imaging—a brilliant application of a phenomenon in physics known as nuclear magnetic resonance, which allows you to identify and map the mass of different atomic nuclei in a volume. The machine was originally called NMRI, but "nuclear" is one of the two forbidden "n" words of our times. So it was dropped from the abbreviation, lest people think they were being lethally irradiated during the measurements. Physicists Felix Bloch and Edward Purcell shared the 1952 Nobel Prize[6] for this discovery. Purcell also happened to be one of my college physics professors. He was into astrophysics and made seminal discoveries related to the behavior of hydrogen atoms,[7] empowering those who use radio telescopes to find and track the existence of massive clouds of hydrogen gas in the Milky Way Galaxy.

A highly valued machine in hospitals, magnetic resonance imaging has no taproots within the profession. No amount of money given to medical researchers would have fueled the discovery of the machine's foundational principles. That's because the MRI is based on laws of physics, discovered by a physicist-stargazer, who had no interest in medicine. The same is true for a hospital's entire radiology department (including X-rays, CT scans, and PET scans), EEGs, EKGs, oximeters, and ultrasound. You name it. If the hospital machine has an on/off switch, its function is probably based on a principle of physics. That's how it works. That's how it has always worked. For the machine to exist at all requires medical engineers who see the utility of such a discovery. The clarion call to fund practical research over fundamental research, and the persistent plea to not spend money in space when we could spend it on Earth, represent noble but underinformed desires. Want to advance civilization? Fund it all. You never know beforehand which discoveries will transform your field, birthed in professions not your own.[8]

Ultrasound technology in particular has contributed to a perennially debated topic regarding the human body. For nearly five months of a nine-month pregnancy, the human fetus cannot survive outside of the womb, even with intensive medical attention. Maybe one day we will know how to bring a fertilized egg to maturity in a medical vessel, but right now, that time feels far off in our future. In the US, arguments rage over how much control we grant to state and federal lawmakers over the uteruses of its citizens. Some demographics feel strongly that pregnant people should not have the right to terminate their pregnancy after the first six weeks, around the time you can first detect a heartbeat via ultrasound.[9] They cite murder.

To be clear, this would be the murder of a nonviable human embryo that weighs no more than a paper clip. Part the curtains of this community and you find strong influence from fundamentalist and otherwise conservative Christian groups. Of the fifteen most religious states,[10] eleven had laws on the books ready to ban or greatly restrict abortion[11] the moment the Supreme Court overturned *Roe v. Wade*, the landmark 1973 case that decriminalized abortion. Clearly, belief in a loving, compassionate Christian God and in the sanctity of all human life (viable or not) strongly motivates these views. They're not being bad citizens, they're being good Christians—although ten of those same eleven states also embrace the death penalty.[12]

Overall, three out of four Republican voters[13] support some kind of anti-abortion/pro-life posture, strictly enforced by laws, in spite of Republicans otherwise wanting less, not more, government in our lives. Medically speaking, for the first eight or nine weeks after conception, an unborn human is an embryo, and a fetus thereafter until birth.[14] In my experience, those who welcome their pregnancy will instead think of the "baby" in their wombs. This simple change in vocabulary emboldens conservative pro-lifers to caricature liberal pro-choicers as simultaneously wanting to "save the whales" and "abort the babies."[15]

Let's look at recent abortion rates in the US. Of the more than 5 million known pregnancies per year between 1990 and 2019,[16] nearly 13 percent were medically aborted.[17] Yet all by itself the uterus spontaneously aborts as many as 15 percent of all known pregnancies during the first twenty weeks. Many more miscarriages go unnoticed since they occur in the first trimester, often before you know you're pregnant. Combined, the number of spontaneous abortions may surpass 30 percent of all pregnancies.[18] So, if God is in charge, then God aborts more fetuses than medical doctors do.

Just some perspectives to consider as we all take sides on who controls our bodies.

How important is our physiology to our livelihood and well-being? Would you be disadvantaged if some of your physiology malfunctioned or neverfunctioned? What does it mean to be handicapped or disabled at all? Dictionaries tell us that your condition markedly restricts your ability to function physically, mentally, or socially. Perhaps model humans exist somewhere—like model homes—and everything about them works perfectly. We could line up, waiting in turn to compare ourselves and decide if any one of us is lacking. These physical specimens would have working fingers, hands, arms, legs, and feet as well as acute senses. They're average height, and all their organs function just as our medical textbooks prescribe. And nothing clouds or diminishes their mental states. This exercise smacks of sensory and physiological chauvinism.[19] Whatever is the ideal, you're not it. Entire industries exist to make you feel inadequate, requiring you to invest boundless time or money or both to achieve normal status. It runs deep. So deep we can hardly think in any other way, but let's try.

Arguably the most famous composition in the canon of classical music is Ludwig van Beethoven's Ninth Symphony, finished in 1824, when he was completely deaf. Was Beethoven disabled? He could hear for most of his life—into his forties—so maybe that's not a good example.

How about a letter written on April 10, 1930, to Captain von Beck of the US Lines' SS *President Roosevelt*. The captain had given a tour of the bridge to a passenger, who later that day waxed poetic about the experience:[20]

> *Again, I stood with the Captain on the bridge, and he was quiet and composed in the presence of "a million universes"—a man with the power of a god. . . . In imagination I saw the Captain standing on the bridge, gazing into the wide-canopied heavens and seeing the darkness sprinkled with stars, systems, and galaxies.*

That passenger was Helen Keller, a 1904 graduate of Radcliffe College, deaf and blind since age nineteen months.

Was Helen Keller disabled?

Matt Stutzman is a championship archer who can outshoot most people who have ever wielded a bow and arrow in competition. He's also a car enthusiast. Oh, and he was born without arms. He shoots his arrows (and fixes his cars) using his uncommonly nimble legs, feet, and toes.[21]

Is Matt Stutzman disabled?

Jahmani Swanson[22] loves basketball, but he wasn't tall enough to play in the NBA, where players average six feet six inches. Jahmani nonetheless continued to work hard at the game, getting better and better. In 2017 he would be discovered by the world-famous Harlem Globetrotters—a team he has played for ever since. His nickname with them is "Hot Shot" Swanson. Hot Shot is a fully grown man, four feet five inches tall, born with dwarfism, a genetic condition that stunts the growth of your long bones. He's one of the most popular players on the Harlem Globetrotters.

Is Hot Shot disabled?

Temple Grandin doesn't think like most other humans. In fact, she thinks more the way farm animals do than their farmers. This curious fact led to a life in animal husbandry, culminating in a doctoral degree in animal science from the University of Illinois. Author of more than sixty research papers and a dozen books, in 2010 she was one of the "Time 100" most influential people in the world.[23] At age two, her delayed development had led to a formal diagnosis of "brain damaged." Temple Grandin has autism, a neurological condition that, in her case, informs and nurtures her unique insights into the minds of farm animals.

Is Temple Grandin disabled?

For most of his professional life, physicist Stephen Hawking had no use of his body. He became paralyzed from a slow, early-onset version of amyotrophic lateral sclerosis—better known as ALS or Lou Gehrig's disease. Meanwhile, his brain still worked, and he made major discoveries in the quantum physics of black holes and cosmology. In 1988 he also wrote *A Brief History of Time*, the single-biggest-selling science book of all time. Helped by machines to assist his reading and writing, Hawking continued to publish and wield a sharp sense of humor his entire life, until his death in 2018.[24]

Was Stephen Hawking disabled?

Oliver Sacks was a noted neurologist, pioneering entire subfields within his profession. He was also a best-selling author, describing the human brain as the "most incredible thing in the universe." He led a remarkably varied life while suffering from a neurological affliction called prosopagnosia, more commonly known as "face blindness." This condition contributed to his severe shyness since he couldn't recognize faces, even if he recognized everything else about you. At times he would even not recognize his own face in the mirror.[25] In 2012, after a lecture on hallucination at Cooper Union college in New York City, I asked him, "If you could go back in time, would you take a magic pill in your youth to cure your neurological disorder?" Without hesitation, he replied "No." His entire professional interest in the human mind was inspired by the very disorders in his own brain. He wouldn't have it any other way.

Was Oliver Sacks disabled?

Jim Abbott wanted his whole life to be a professional baseball player—a dream shared by many American boys. Jim wanted to be a pitcher in the major leagues. He succeeded and played for many teams, chalking up a mixed record of wins and losses. But on September 4, 1993, while playing for the storied New York Yankees, he pitched a no-hitter—that's when no batter gets a hit in the entire game. There have been about 320 no-hitters in major league history out of 220,000 games played. Due to a congenital birth defect, Jim Abbott was born without a right hand.

Is Jim Abbott disabled?

Ludwig van Beethoven, Helen Keller, Temple Grandin, Stephen Hawking, and Oliver Sacks all had feature-length movies made of their lives, with marquee actors portraying them. Matt Stutzman, Hot Shot, and Jim Abbott are surely next. Every one of them is (or was) better at what they did professionally than nearly all other humans on Earth.

Perhaps they all achieved, not despite their disability but because of it.

This concept knows no bounds. For example, if you don't need a curb cut to reach the sidewalk from the street, but a person in a wheelchair does, yet the person in the wheelchair knows vector calculus and you don't, shall we classify your mathematical illiteracy as a disability? Is a student developmentally delayed, or is the teacher simply incompetent?

Neil deGrasse Tyson
@neiltyson

Some educators who are quick to say, "These students just don't want to learn" should instead say to themselves, "Maybe I suck at my job."

6:27 PM · Mar 16, 2022 · Twitter Web App

Are non-artists disabled because they can't draw? When you are not good at one thing, you typically try something else. In a free society, there's lots of "something elses" out there. Even better, just do what you love, regardless of others' efforts to dissuade you. You may just succeed famously in the face of naysayers who seek to standardize who should or shouldn't succeed at all. Which brings a modern proverb to mind:

> *Those who were seen dancing were thought to be insane*
> *by those who could not hear the music.*[26]

Maybe everybody is disabled in some way. If so, that means nobody is disabled.

As a species, how does our mind measure up? Most of us have embraced the notion that we use only 10 percent of our brain. This dates back more than a century, but was never true.[27] It nonetheless persists because deep down it serves our longings. Psychics want it to be true so that they can claim untapped powers of mind await us all. Teachers want it to be true so they can motivate their underperforming students. The rest of us want it to be true because it gives us hope for ourselves. Brain scans reveal that we engage much more than 10 percent, but that some fraction of our brain never lights up at all, no matter the stimulus[28]—the neurological equivalent of cosmic dark matter.

Humans are no doubt the smartest creatures ever to exist in Earth's tree of life. Our brain consumes 20 percent of our body's energy,[29] so even our physiology values the organ. In our search for extraterrestrial intelligence, we presume they're at least as smart as we are. Yet a few simple facts should force us all to take pause. Who assessed humans as intelligent? Humans did. There's that hubris again—an ego boosting itself. Let's continue. We are vastly more intelligent than the next smartest species of life on Earth—the chimpanzee. Yet we share 98+ percent identical DNA with them. What a difference that 2 percent makes! We have poetry and philosophy and art and space telescopes. Whereas the smartest chimp might stack boxes to reach a banana suspended from above—something human toddlers can do. Or they might select the right kind of twig for extracting tasty termites from a mound. So how can that (small) 2 percent difference account for what we declare to be our vast intelligence relative to chimps?

Maybe the difference in our respective intelligences is as small as that 2 percent difference in DNA would indicate. This thought doesn't occur to any of us because of how invested we are in distinguishing our place in the animal kingdom. The tree of life is packed with animals that do things better than us. In other words, if the Summer Olympics were open to all species of animals, we would lose practically every event. That "faster, higher, stronger" motto leaves humans far behind in the animal kingdom.

There's one thing we're physically better at than all other animals. We can stalk any land animal to exhaustion. Cave paintings of early humans commonly depict hunters of deer, bison, and other large grazing mammals, including mammoths. Each species is stronger than us and can outrun us, but they can't run forever. Our mostly hairless bodies allow us to sweat efficiently and cool as we pursue our dinner, while our furry prey eventually overheat and collapse. Spears would help too, shortening the chase. This tactic works marvelously, so long as your prey are herbivores. If you stalk a carnivore, it could just turn around, stalk back, and eat you instead. One might suspect that any cavemen who craved lion meat were rapidly withdrawn from the gene pool.

Sweaty bodies notwithstanding, our best asset is our brain. Yes, we have huge mammalian brains, but they're not the hugest. Whales, elephants, and dolphins each have brains larger than ours. That's not good for our ego. Let's keep trying. How about dividing brain weight by body weight, forming a brain-to-body-weight ratio. Ahh. That's better. A ranking of all mammals by that metric puts none higher than us, allowing us to feel good about something.

With all this brain-gerrymandering, unwelcome anomalies arise.[30] For example, the brain-to-body-weight ratio in mice rivals that of humans, so we don't dominate this list. If we open the ranking to include all vertebrates, not just mammals, then we lose to small birds like parrots and medium birds like crows. There's even a YouTube video showing a magpie drinking water from a bottle in the street,[31] but with limited reach of its beak down to the water level. After each sip, the level drops and the magpie searches for and finds a rock that will fit through the neck of the bottle. The bird drops it in, thereby raising the water level so that it can keep drinking. In the video, the magpie repeats this Archimedean feat seven times. Over centuries and millennia, one thing is for sure, we've persistently underestimated the intelligence of our fellow animals—further evidence of our fragile ego—only to be shocked when they do something smart.

If we open the brain-to-body-weight contest to all animals, not just vertebrates, then ants win spectacularly. On average, the human brain is 2.5 percent of our body weight, yet for some ant species, their brains are closer to 15 percent of their body weight. Thinking this through, we're forced to conclude that visiting space aliens who prioritize brains might first try to chat with the ants, then the birds, then perhaps the whales, the elephants, the dolphins. Then the mice. And then maybe—just maybe—the humans.

Embarrassing.

But we can figure things out with our impressive intelligence, and build stuff using our opposable thumbs. No doubt we're top of the heap in those categories, which returns us to the chimp-human comparison. Imagine alien life with 2 percent different DNA from us along the same intelligence vector that distinguishes us from chimps. On that scale, if the smartest chimps can do what human toddlers can do, then the smartest humans could do what this alien life-form's toddlers can do. Actual aliens might not have DNA at all, but that doesn't change the thought experiment. What if their version of our primatologists search for the smartest human on Earth? They might have found Stephen Hawking before he died. If so, they would just roll him forward at their scientific conference and declare that the good Professor can do astrophysics calculations in his head, just like little Zadok Jr. returning home from preschool. Zadok proceeds to show a derivation of the fundamental theorem of calculus to its parents. They reply, "Awww, that's cute. Let's get a magnet and display it on the refrigerator door!"

The simplest thoughts of these adult aliens would land beyond human comprehension. This sentence alone, "Let's meet at 10:30 a.m. for a cup of coffee and discuss the quarterly report before it goes public with the press," contains a half-dozen concepts incomprehensible to a chimp. Consider further that no matter how bad you were at long division, you were much better at it than any chimp will ever be. Given these stark realities, our alien 2 percenters might not rate humans as intelligent at all. Just imagine the thoughts and discoveries and inventions they'd be capable of. Actually, we can't. We literally can't. For them, there's only a trifling difference between stacking boxes to reach bananas and the design and launch of space telescopes. As for life-forms that possess 5 or 10 percent more smarts than humans, the bigger that percentage gets, the more we will appear to them as worms appear to us.

It gets worse.

Beyond simple hand gestures, we don't know how to communicate meaningfully with chimpanzees. We can't even tell them, "Come back tomorrow afternoon. I've got a new shipment of bananas arriving for you." Assessing the effort that we invest trying to get big-brained mammals to do what we say, we tend to measure their intelligence by an ability to understand us, rather than measure our intelligence by an ability to understand them. Since we can't meaningfully communicate with any other species of life on Earth—not even those genetically closest to us—how audacious of us to think we can converse at all with intelligent alien life upon first meeting them.

Cosmic perspectives wield the power to humble our human hubris, with full justification for doing so. But the question remains, do we have a seat at the table among the intelligent life-forms of the universe? Do we possess sufficient smarts to answer the cosmic questions we've posed? Do we possess sufficient intelligence to even know what questions to ask?

Where does that leave us? Can the mind ever understand how the brain works? By the same measure, can the universe create something more complex, more capable, more better than the universe itself? I lose sleep over that thought, and it's not due to the questionable grammar. Maybe so, because of the following example. We often admire the complexity of our human brain: that the number of neurons it contains rivals the number of stars in the Milky Way Galaxy;[32] that we possess amazing powers of rationality and thought; that our frontal lobe confers high-level, abstract reasoning. Yet we've built computers that outthink us in practically every brain contest we have ever conceived for ourselves. You know the list. It's long and humiliating. Now add to it a mechanical computer that can solve the famed Rubik's Cube in 0.25 seconds.[33] Not that I'm any measure of this but, having never read any solutions, my personal best from long ago is 76 seconds. That's three hundred times slower than the machine. Computers will soon be driving all our cars—faster, more efficiently, and with ever-fewer traffic fatalities, instead of the thirty-six thousand deaths we currently endure annually in the US and 1.3 million worldwide.[34] So whether or not the universe can make something more complex than itself, we've managed to make something more capable than ourselves, and we've only just begun.

If a wafer of silicon with some electrical current passing through it is capable of outperforming our brains in so many ways, maybe we've overvalued our capacity for thought. Not surprising. We like thinking highly of ourselves. Consider further that full-grown, educated people walk among us who fear the number 13, who are sure Earth is flat, and who blame unfortunate events in their lives on the planet Mercury being in retrograde. Not only that, simple chemicals introduced to (or removed from) the brain greatly disrupt our perceptions of objective reality. If not now, then soon, as computing power continues to improve, we will surely simulate a far more rational version of ourselves in a video universe than we could ever live up to—all the good and none of the bad features of being human.

One might ask if we're living in such a simulation now? The reasoning goes: Intelligent life in the one real universe evolves and invents powerful computers to program intelligent life that is so real, the life-forms themselves feel a sense of free will and have no clue they're simulated. They evolve enough to invent powerful computers of their own, and program life that is so real, they don't know they're simulated either. Continue this exercise for as long as you like. If you cover your eyes and throw a dart, you are far more likely to hit a simulated universe than the original real one that started it all. Hence, we're likely living in a simulation. Yes, that's how the reasoning goes. On reflection, however, a purposefully conceived computer would not likely be capable of all the inane and irrational behavior that we've exhibited in the history of our species. Computers are better than that. So this could be the best evidence yet that we don't live in a simulation. Call it the inanity defense.

CODA

Life & Death

Viewed with naive eyes, to watch a fetus grow inside a womb and spring forth from the loins of another human is an event befitting space aliens and their imagined physiology. Unless you're an obstetrician or a midwife, human birth is one of the rarest common things you will ever witness—worldwide, on average, four babies are born every second.[1] With every birth, a new consciousness has entered the world with an ever-increasing life expectancy, thanks to the advances of modern medicine. Another rare, common thing is death. Unless you're an emergency room nurse, the medical examiner, or an active-duty soldier in armed conflict, you might witness the death of only three or four humans your entire life. Yet worldwide, sixty million people die each year. That's about two per second, on average.[2]

We can expect to live twice as long today as people in the year 1900.[3] Walk the rows of any old cemetery and do the math. The chiseled birth and death dates on each tombstone bear silent witness to the shortened life expectancies of bygone eras. You'll be glad you live today and not any time in the past. But in one or two hundred years, will those who tour the cemeteries that contain all our remains think the same of us, as they pity our paltry eighty-year life expectancies? Will they be the ones who live long enough to travel among the distant planets and the stars themselves?

Suppose we could live forever.

It's better to be alive than dead. Though more often than not, we take being alive for granted. The question remains, if you could live forever, would you? To live forever is to have all the time in the world to do anything you ever wanted. You can even mutiny a generation starship and return home to Earth if you wanted to. Seems like an attractive idea, but maybe the knowledge of death creates the focus that we bring to being alive. If you live forever, then what's the hurry? Why do today what you can put off until tomorrow. There is perhaps no greater de-motivating force than the knowledge you will live forever. If true, then knowledge of your mortality may also be a force unto itself—the urge to achieve, and the need to express love and affection now, not later. Mathematically, if death gives meaning to life, then to live forever is to live a life with no meaning at all.

For these reasons death may be more important to our state of mind than we are willing to recognize. If you were to bring a colorful bouquet of flowers to a loved one, and those flowers were made of plastic, or even silk, they will surely be less appreciated than if they were real. Flowers that live forever miss the point. We seek the increasing beauty of each flower in a bouquet, as they unfurl one by one in the light of the day.

We're absorbed by their irresistible aromas. We duly accept the care and feeding they require. We embrace their senescence as the stems weaken, no longer sustaining the weight of the faded petals. Florists remain in business because the death of flowers—typically within a week of receiving them—is precisely what gives them meaning to your loved one. Compare that with forever-flowers that require no maintenance, never die, have no smell, yet remain just as beautiful a week, a month, or a year later. They even gather dust.

Dogs aren't flowers, but they convey a similar story. Ever notice how enthusiastic they can be? If you let them, dogs will jump all over you and lick your face. They'll chase and retrieve things you throw. They're ecstatic when you return home, even if you just went out to the mailbox and came right back. They love every minute you spend with them. For most dogs, each day matters. Humans live approximately seven times longer than dogs.[4] That's the origin of the famous "dog years" calculation. Multiply your dog's actual age by seven, and you get an equivalent age for a human. Keeping with that seven-to-one ratio, a day unto a dog is a week unto humans. Maybe that's why they make every day count. Like flowers on the mantel, not a day goes by when they do not compel you to take notice, and smile. If your family brought home a puppy dog during your childhood years, you saw it grow and eventually get old and die while you were either in high school or college. You surely remember those years.

Not everything dies for having grown old. Contrary to the collective delusion that Mother Nature is a nurturing, caring entity that cradles and protects all its forms of life, Earth is instead a giant killing machine. Holding aside all the climatic and geologic forces that would just as soon have you dead, such as droughts, floods, hurricanes, tornadoes, earthquakes, tsunamis, and volcanoes, there's no end of creatures that want to suck your blood, inject you with venom, infect your physiology, or simply eat you.

The universe wants to kill you too.

At least one of the six extinction episodes in Earth's timeline of life—the Cretaceous-Tertiary (K-T) event[5] from 66 million years ago—was partly or entirely triggered by the impact of a rogue asteroid the size of Mount Everest.[6] With no space program at the time to deflect the impactor, it was a bad day for the dinosaurs. Was also a bad day for 70 percent of all species on land and in the oceans. They too went extinct. And if you think that was bad, during the so-called Permian-Triassic extinction of 250 million years ago, life on Earth almost ended entirely.[7]

Modern humans are complicit in Mother Nature's wrath. Our encroachment on pristine ecosystems is rendering species extinct at a rate up to a thousand times the ongoing level that it naturally occurs.[8] Geologists have named a stretch of time to identify our upheaval of Earth's biosphere: from the dawn of agriculture 11,700 years ago through today, they call the Holocene Epoch.

Of all species that ever lived on Earth, 99.9 percent of them have gone extinct.[9] Who knows what wonders of biodiversity died out of the world for want of the luck, the strength, or the will to survive?

Suppose we did live forever. One practical problem: if everyone born never dies, and if people keep making babies, then Earth's population will rapidly outstrip the resources to support it. So the day we stop dying must also be the day we find another orb to accommodate our overproduction of air-breathing humans. This need for extra planets will never cease. But the universe is vast, and just in our small sector of the Galaxy, the catalogues are now rising through five thousand known exoplanets. We just need to invent terraforming technologies and either warp drives or wormhole transportation systems, and all will be fine.

We want to live forever because we fear death. We fear death because we are born knowing only life. Yet we don't fear having never been born. While it's surely better to be alive than dead, it's even better to be alive at all than to not ever exist. Religion through the ages has offered detailed accounts for what happens after death. For some, it includes what happened before you were born—a basic tenet of reincarnation. Science doesn't have much to say about Valhalla or Elysium or Hades or Heaven or Hell or the spirits of your ancestors. The methods and tools of science do however make cold, concrete statements about what happens when you die. You spend your life eating food, which, among other things, delivers calories to your body. A calorie is simply a unit of energy. Your body makes heat from these calories, maintaining your body temperature at nearly 100 degrees Fahrenheit even though nothing else around you gets that hot. Biologically, humans need to be at that temperature to function. You also need energy to walk and talk and do things while you're awake. You also need energy when you're doing nothing. These are the primary reasons why we eat at all.

The moment you die, you stop metabolizing your calories, and your body slowly drops to room temperature. At a funeral, if you touch the person in the casket, typically the exposed hand of a folded arm, you instantly realize that the body is cold. Even at room temperature, the body feels cold when compared with the hand of a living person who's still burning energy.

Most biological molecules harbor energy. When someone gets cremated—when the molecules burn—that energy escapes in the form of heat, which warms the air of the crematorium's chimney, which then radiates as infrared photons into Earth's atmosphere and eventually into space, traveling at the speed of light. Sounds morbidly romantic, but when I die, I'd rather be buried. My infrared energy crossing the vacuum of space is of no use to anybody or anything at any time. Put me in the ground, six feet under, and let the worms and microbes crawl in and out of my carcass as they dine upon my flesh. Let the root systems of the plant and fungal kingdoms of life extract nutrients from my body. The energy content of my molecules, which I had assembled throughout my lifetime by consuming the flora and fauna of this Earth, will return to them, continuing the biosphere's cycle of life.

Upon death, there's no evidence that you experience the consciousness you enjoyed while alive. The electrochemical source of all your thoughts, feelings, and sensory awareness of the universe—your brain, which normally lights up an MRI—becomes starved of oxygen. We know that's you disappearing because people who experience a series of ultimately fatal strokes tragically and systematically lose function of their mind and body as they descend into a state of nonexistence. That's not as odd as it sounds. Were you conscious before you were conceived? Did you complain, "Where am I? How come I'm not on Earth?" No, you simply didn't exist, and if you're lucky to be born, your nonexistence before life bookends your nonexistence after death.

~

Apart from whatever priceless value religions might assign to life, economists have no moral hesitation calculating what you're worth dead. Tort-law courts have been doing it for years, using many approaches.[10] The simplest calculation estimates your future earned income if, from the negligence of others, you lost your life or became permanently disabled and could no longer sustain your livelihood.

Another sample calculation[11] comes from enticing people to work a dangerous job by paying them more than an equivalent job that carries no risk of life. If you want an extra $400 per year to take a job where there's a 1 in 25,000 chance of dying in that year then, whether or not you know it, you have personally valued your own life at $400 x 25,000 = $10 million.

In a different kind of calculation, we can instead assess your debt to civilization. From birth until your first full-time job, either after high school or college, your family, your city, your state, and your country have been investing in you. Every child needs food and diapers and shelter. Depending on how privileged you are—if you've had expensive day care or nanny care or tutoring or private schools—this can sum to upwards of a million dollars per child. So let's just look at an estimate for a middle-income family with two children. That's $233,000[12] to raise a child from birth to age eighteen. Attend four years of state college and add another $100,000. Private college, double it. At that moment, if you died, hundreds of thousands of dollars invested by others in you and your future evaporates instantly. All opportunities for a return on that investment are gone. Yet that is precisely the age when you are drafted into the armed services. Of the 58,000 American deaths in the Vietnam War, the last war to conscript soldiers, 61 percent were twenty-one or younger.[13] That's 35,000 people, killed at the exact moment that would otherwise begin their economic payback to the economy. Military hawks would say their life in battle is the ultimate payback to the country. If it's better to be alive than dead, then the ultimate payback would instead be to do whatever it takes to not kill one another for harboring different views of the world, ensuring a long and healthy life for everyone.

Consider that humans are typically conceived in the single most intimate act of human affection. We then gestate in utero for nine months, suckle for another twelve months, and require continual care through our toddler years. Afterward, humans attend elementary school to learn reading, writing, and arithmetic. In middle school and high school we also learn biology, chemistry, and maybe physics. We read works of literature. We learn history and the arts and might even play sports. Lifelong friendships germinate from these activities. We may also learn languages spoken by other humans around the world. We participate in all the seasonal rituals that we retain in modern society as a binding force that brings us together. Adulthood arrives. Twenty-one years go by. At a speed of 30 kilometers per second, Earth completed twenty-one orbits around the Sun—a total of 20 billion kilometers through space.

All along, humans invent, refine, and perfect antipersonnel weapons such as land mines, assault rifles, missiles, and bombs, any one of which can end a life in an instant. How long is an instant? A bullet from a high muzzle velocity rifle, moving at three times the speed of sound, can pass through your chest, rip a hole in your heart, and emerge from your back in less than four ten-thousandths of a second—before you even heard the weapon fire. For a bomb of any size, its shock wave does most of the damage. The force of rapidly expanding air, upon reaching where you stand, can blow apart your body in a thousandth of a second. One can also die prematurely by accident or from an unfortunate disease, but we invented these weapons with the sole purpose of killing other humans—our own kind—in an instant. Wars have taken a staggering and tragic toll on human life since pre-civilization. Yet even excluding organized armed conflict, humans find reasons to murder other humans at a rate that exceeds 400,000 per year—yes, worldwide, humans commit homicide more than a thousand times a day.

In spite of all this carnage, humans are not the deadliest animal against humans. That distinction goes not to lions or tigers or bears. Nor to snakes or sharks. It goes to mosquitoes, carriers of the viral infections yellow fever, dengue, and especially the parasite malaria, itself killing more than a half million people per year,[14] mostly young children. Lethal Mother Nature, at it again.

~

Do you know—do you really know—how precious life is?

The total number of people who have ever been born is about 100 billion. Yet the genetic code that generates viable versions of us is capable of at least 10^{30} variations.[15] That astronomically huge number is a one followed by 30 zeros, providing a million-trillion-trillion possible souls.

Run through them all and we end up with you again—or at least your twin. But that won't happen any time soon. So far, our branch of the tree of life has produced no more than 0.00000000000000001 percent of all possible humans, forcing the conclusion that most people who could ever exist will never even be conceived.[16] Each of us, for all practical purposes, is unique in the universe—now and forever.

Being alive is the time to celebrate being alive—every waking moment. Along the way, why not strive to make the world a better place today than yesterday, simply for the privilege of having lived in it. On my deathbed, I'd be sad to miss the clever inventions and discoveries that arise from our collective human ingenuity, presuming the systems that foster such advances remain intact. That's what fueled the exponential growth of science and technology in my lifetime. I'd further wonder whether civilization's arc of social progress will continue—with all its fits and starts—and thus reward any time traveler from the oppressed spectrum of humanity who chose to visit the future rather than the past.

On death, I'd miss out on the adult lives of my children. But no tragedy there. Just a selfish yearning against something that is natural and normal. I'm supposed to predecease them. The real tragedies are when your children die before you do. Something Gold Star families know all too well, with sons and daughters and brothers and sisters lost to wars.

On the whole, I don't fear death. Instead, I fear a life where I could have accomplished more. An epitaph worthy of a tombstone comes from the nineteenth-century educator Horace Mann:[17]

I beseech you to treasure up in your hearts these my parting words.

Be ashamed to die until you have won some victory for humanity.

Our primal urge to keep looking up is surely greater than our primal urge to keep killing one another. If so, then human curiosity and wonder, the twin chariots of cosmic discovery, will ensure that starry messages continue to arrive. These insights compel us, for our short time on Earth, to become better shepherds of our own civilization. Yes, life is better than death. Life is also better than having never been born. But each of us is alive against stupendous odds. We won the lottery—only once. We get to invoke our faculties of reason to figure out how the world works. But we also get to smell the flowers. We get to bask in divine sunsets and sunrises, and gaze deeply into the night skies they cradle. We get to live, and ultimately die, in this glorious universe.

ACKNOWLEDGMENTS

For a book of this scope, I am grateful to many people who read early (and late) drafts and offered comments that derive from their passions, interests, and expertise: my AMNH museum colleagues Ian Tattersall (paleoanthropologist) and Steven Soter (astrophysicist) provided valued insights that deepened my treatment of multiple topics. Friends Gregg Borri (attorney and military historian), Jeff Kovach (investor), Paul Gamble (former navy JAG and judge), Ed Conrad (conservative economist), Erin Isikoff (medieval novelist), Heather Berlin (neuroscientist), and Irwin Redlener (MD and public health activist) each offered expertise that benefited related topics in this book. I am privileged to have such an array of professionals within arm's reach. Magatte Wade (culture entrepreneur) and Nicholas Christakis (sociologist), both new acquaintances, also provided insights that informed and enhanced several of the arguments presented.

Brother Stephen Tyson Sr. (artist and philosopher) weighed in on truth and beauty as any good artist would. Daughter Miranda Tyson (social justice educator) and son Travis Tyson (college senior) both made it clear that I may never be woke enough for their outlooks on the world, which helped bring many passages of the book into the third decade of the twenty-first century, where they belong. Sister-in-law Gretchel Hathaway (DEI specialist) offered a few woke comments of her own. Cousin Greg Springer (libertarian south Texas rancher) helped tighten points of view that I didn't know needed tightening. Niece Lauryn Vosburgh (wellness consultant) is a New Age rationalist and my conduit to that world of thought. Tamsen Maloy (lapsed Mormon and lifelong vegetarian) helped me flesh out various topics that I had otherwise skimmed in my treatment. Comedian Chuck Nice offered advice on occasional turns-of-phrase to boost the chances of the reader cracking a smile. And wife Alice Young (mathematical physicist) is wise on all fronts, offering helpful and insightful comments where I most needed them and where I least expected them.

I'm also thankful to the Dutch language translator for *Starry Messenger*, Jan Willem Nienhuys, who knows my work from previous books. He's deeply fluent in math, physics, and astronomy, and manages to catch errors of omission and commission that slipped by me and a dozen other experts.

Not all analysis is served by internet search engines. Some seemingly simple quantities offered herein derive from extensive data compilations. For this, I thank researcher Leslie Mullen for her ability to distill vibrant knowledge from lifeless data.

I'm also thankful to friend Rick Armstrong for connecting me with Kimberly Mitchell, daughter of Edgar Mitchell

(*Apollo 14*). Delighted to now count her as a friend, who is no less committed to doing right by the world than her father was. Edgar Mitchell's beginning quote for *Starry Messenger* opens the portal through which the entire book flows.

I continue to be delighted by the enthusiasm that Henry Holt & Company has shown for this project, especially via Tim Duggan (executive editor), Sarah Crichton (editor-in-chief), and Amy Einhorn (publisher). What I saw for the book and what they imagined for it were highly resonant.

Lastly but not leastly, Betsy Lerner (poet, literary agent) has supported my writings for thirty years. She helped nip-tuck the content of all chapters in this book, advising on tone and flow while invoking her highly literate sensibilities all along the way.

CREDITS

Edgar Mitchell quote: used with permission from Kimberly Mitchell.

Mike Massimino "Heaven" quote: used with permission.

Theodore Roosevelt Equestrian Statue at the American Museum of Natural History: ©AMNH/Denis Finnin.

Hand of God: ©NASA Orbiting Nuclear Spectroscopic Telescope Array (NuSTAR).

Physicists in Town: author photo of APS News headline, https://www.aps.org/publications/apsnews/199908/knowledge.cfm.

"*They're Made out of Meat*" excerpt: used with permission from Terry Bisson.

Tweets: original Tweets ©Neil deGrasse Tyson, via Twitter.com.

Vanity Card No. 536, excerpt: used with permission from Chuck Lorre.

NOTES

DEDICATION

1. "memory of Cyril DeGrasse Tyson": Joseph P. Fried, "Cyril D. Tyson Dies at 89: Fought Poverty in a Turbulent Era," *New York Times*, December 30, 2016, accessed January 19, 2022, https://www.nytimes.com/2016/12/30/nyregion/cyril-degrasse-tyson-dead.html.

OVERTURE: SCIENCE & SOCIETY

1. "survival advantages": Michael Shermer, *The Believing Brain* (New York: Times Books, 2011).
2. "prejudice or leniency": A. I. Sabra, ed., *The Optics of Ibn al-Haytham, Books I–III: On Direct Vision*, Arabic text, edited and with introduction, Arabic-Latin glossaries, and concordance tables (Kuwait: National Council for Culture, Arts and Letters, 1983).
3. "their own opinion": *The Notebooks of Leonardo da Vinci*, vol. 2, trans. John Paul Richter, chapter XIX: Philosophical Maxims. Morals. Polemics and Speculations. II. Morals; On Foolishness and Ignorance. Maxim no. 1180 (New York: Dover, 1970), 283–311, accessed March 19, 2022, https://en.m.wikisource.org/wiki/The_Notebooks_of_Leonardo_Da_Vinci/XIX.

CHAPTER ONE: TRUTH & BEAUTY

1. "Ode on a Grecian Urn": John Keats, "Ode on a Grecian Urn," accessed March 1, 2022, https://www.poetryfoundation.org/poems/44477/ode-on-a-grecian-urn.
2. "God and Truth": John 14:6, King James Version.
3. "Vincent van Gogh": Date and time established by the author from the phase of the Moon, its orientation, and elevation on the sky relative to Venus.
4. "Sermon on the Mount": Clifford M. Yeary, "God Speaks to Us on Tops of Mountains," Catholic Diocese of Little Rock (website), April 26, 2014, accessed October 30, 2021, https://www.dolr.org/article/god-speaks-us-tops-mountains.
5. "Mesoamerican pyramids": Dave Roos, "Human Sacrifice: Why the Aztecs Practiced This Gory Ritual," History, October 11, 2018, accessed October 30, 2021, https://www.history.com/news/aztec-human-sacrifice-religion.
6. "cloud nine": Paul Simons, "The Origin of Cloud 9," *The Times* (London), September 6, 2016, accessed October 30, 2021, https://www.thetimes.co.uk/article/weather-eye-7ftq5tvd2.
7. "catalog IDs": The field of astrophysics contains many catalogs of many things. Here we reference: PSR: Pulsating Source of Radio (pulsars); NGC: New General Catalogue of Nebulae and Clusters of Stars; IC: Index Catalogue of Nebulae and Clusters of Stars—an extension of the NGC.
8. "it came from the Moon": *StarTalk Radio*, "Decoding Science and Politics with Bill Clinton," November 6, 2015, accessed October 30, 2021, https://www.startalkradio.net/show/decoding-science-and-politics-with-bill-clinton/.
9. "8.7 million species": National Geographic Society Resource Library, "Biodiversity," accessed October 30, 2021, https://www.nationalgeographic.org/encyclopedia/biodiversity/.
10. "extinct": Hannah Ritchie and Max Roser, "Extinctions," *Our*

World in Data, accessed October 30, 2021, https://ourworldindata.org/extinctions.

11. "ground-to-cloud": Not a typo. It's a little-known fact that the visible stroke of a lightning bolt moves from the ground to the clouds and not vice versa.
12. "Kilmer's most famous poem": Joyce Kilmer, "Trees," Oatridge, accessed November 2, 2021, https://www.oatridge.co.uk/poems/j/joyce-kilmer-trees.php.

CHAPTER TWO: EXPLORATION & DISCOVERY

1. "debated in India": Simon Mundy, "India Critics Push Back Against Modi's Space Programme Plans," *Financial Times*, August 27, 2018, accessed July 11, 2021, https://www.ft.com/content/edeb1846-a691-11e8-8ecf-a7ae1beff35b.
2. "live in squalor": "Poverty in India: Facts and Figures on the Daily Struggle for Survival," SOS Children's Villages, accessed July 11, 2021, https://www.soschildrensvillages.ca/news/poverty-in-india-602.
3. "T. S. Eliot once mused": T. S. Eliot, "Little Gidding," *Four Quartets*, 1942, accessed July 8, 2020, http://www.columbia.edu/itc/history/winter/w3206/edit/tseliotlittlegidding.html.
4. "voyage to the Caribbean": "Columbus Reports on His First Voyage, 1493," Gilder Lehrman Institute of American History, accessed July 9, 2021, https://www.gilderlehrman.org/history-resources/spotlight-primary-source/columbus-reports-his-first-voyage-1493.
5. "discovery of ideas": Neil deGrasse Tyson, "Paths to Discovery," in *The Columbia History of the 20th Century*, edited by Richard Bulliet (New York: Columbia University Press, 2000), 461.
6. "large collaborations": Daniele Fanelli and Vincent Larivière, "Researchers' Individual Publication Rate Has Not Increased in

a Century," *PLOS ONE* 11, no. 3 (March 9, 2016), accessed July 10, 2021, https://journals.plos.org/plosone/article?id=10.1371/journal.pone.0149504.

7. "Patent and Trademark Office": "US Patent Statistics Chart Calendar Years 1963–2020," US Patent and Trademark Office, accessed July 19, 2021, https://www.uspto.gov/web/offices/ac/ido/oeip/taf/us_stat.htm.
8. "velocipede": "History of the Bicycle: A Timeline," Joukowsky Institute for Archaeology and the Ancient World, Brown University, accessed July 8, 2021, https://www.brown.edu/Departments/Joukowsky_Institute/courses/13things/7083.html.
9. "closed-circuit trip": "Italians Establish Two Flight Marks," *New York Times*, June 3, 1930.
10. "Aero Club of America": Original letter from the library of the author.
11. "nuclear warheads": Hans M. Kristensen and Matt Korda, "Status of World Nuclear Forces," Federation of American Scientists, accessed July 9, 2021, https://fas.org/issues/nuclear-weapons/status-world-nuclear-forces/.
12. "American Psychiatric Association": Neel Burton, "When Homosexuality Stopped Being a Mental Disorder," *Psychology Today*, September 18, 2015, accessed September 8, 2021, https://www.psychologytoday.com/us/blog/hide-and-seek/201509/when-homosexuality-stopped-being-mental-disorder.
13. "civilization doomed": Recounted from personal communication at the Planetary Society's 25th Anniversary Gala, Los Angeles, 2005.
14. "Ecclesiastes": Ecclesiastes 1:9, King James Version.

CHAPTER THREE: EARTH & MOON

1. "Earth from space": Mike Massimino, *Spaceman: An Astronaut's Unlikely Journey to Unlock the Secrets of the Universe* (New York: Crown/Archetype, 2016).

2. "GDP per capita": Worldometer, "GDP per Capita," accessed July 8, 2021, https://www.worldometers.info/gdp/gdp-per-capita/.
3. "*Apollo 8* saved 1968": Alice George, "How Apollo 8 'Saved 1968,'" *Smithsonian*, December 11, 2018, accessed July 6, 2021, https://www.smithsonianmag.com/smithsonian-institution/how-apollo-8-saved-1968–180970991/; Kelli Mars, ed., "Dec. 27, 1968: Apollo 8 Returns from the Moon," NASA, last updated December 27, 2019, accessed July 6, 2021, https://www.nasa.gov/feature/50-years-ago-apollo-8-returns-from-the-moon.
4. "Santa Barbara County": Christine Mai-Duc, "The 1969 Santa Barbara Oil Spill That Changed Oil and Gas Exploration Forever," *Los Angeles Times*, May 20, 2015, accessed March 28, 2022, https://www.latimes.com/local/lanow/la-me-ln-santa-barbara-oil-spill-1969–20150520-htmlstory.html.
5. "Ohio National Guard": Jerry M. Lewis and Thomas R. Hensley, "The May 4 Shootings at Kent State University: The Search for Historical Accuracy," Kent State University, accessed July 6, 2021, https://www.kent.edu/may-4-historical-accuracy.
6. "Earth Day history": "About Us: The History of Earth Day," Earth Day (website), accessed July 6, 2021, https://www.earthday.org/history/.
7. "flag for Earth Day": "Earth Day," Wikipedia, accessed July 6, 2021, https://en.wikipedia.org/wiki/Earth_Day.
8. "polluted cities": Water and Power Associates, "Smog in Early Los Angeles," accessed July 6, 2021, https://waterandpower.org/museum/Smog_in_Early_Los_Angeles.html.
9. "DDT": (1) *Use of Pesticides: A Report of the President's Science Advisory Committee*, President's Science Advisory Committee, May 1963.

   (2) *Restoring the Quality of Our Environment: A Report of the Environmental Pollution Panel President's Science Advisory Committee*, President's Science Advisory Committee, November 1965.

(3) *Report of the Committee on Persistent Pesticides: Division of Biology and Agriculture, National Research Council to the Agriculture Department*, National Research Council, May 1969.

(4) *Report of the Secretary's Commission on Pesticides and Their Relationship to Environmental Health*, US Department of Health, Education, and Welfare, December 1969.

10. "riverbanks": "Mississippi River Oil Spill (1962–63)," Wikipedia, accessed September 16, 2021, https://en.wikipedia.org/wiki/Mississippi_River_oil_spill_(1962–63).
11. "sixth seal": Revelation 6:12, King James Version.
12. "apocalyptic verse": Revelation 6:13, King James Version.
13. "prophecy of doom": Story recounted by Lincoln's friend Walt Whitman, and discussed by Donald W. Olson and Laurie E. Jasinski, "Abe Lincoln and the Leonids," *Sky & Telescope* (November 1999): 34–35, accessed July 24, 2022, https://skyandtelescope.org/wp-content/uploads/LincolnandLeonids.pdf.
14. "*Pale Blue Dot*": Carl Sagan, *Pale Blue Dot: A Vision of the Human Future in Space* (New York: Random House, 1994).

CHAPTER FOUR: CONFLICT & RESOLUTION

1. "no tolerance": For an academic analysis: Eli J. Finkel et al., "Political Sectarianism in America," *Science* 370, no. 6516 (October 30, 2020): 533, accessed December 19, 2021, https://pcl.sites.stanford.edu/sites/g/files/sbiybj22066/files/media/file/finkel-science-political.pdf.
2. "military ideologues": See, e.g., Neil deGrasse Tyson and Avis Lang, *Accessory to War: The Unspoken Alliance Between Astrophysics and the Military* (New York: W. W. Norton, 2018).
3. "V2 ballistic missile": See "Did We Hit the Wrong Planet?," *JF Ptak Science Books* (blog), accessed July 2, 2021, https://longstreet.typepad.com/thesciencebookstore/2013/08/did-we-hit-the-wrong-planet-w-von-braun-1956.html.

4. "London and Antwerp": "V-2 Rocket," Wikipedia, accessed July 2, 2021, https://en.wikipedia.org/wiki/V-2_rocket#Targets.
5. "evolutionarily understandable": See, e.g., K. Jun Tong and William von Hippel, "Sexual Selection, History, and the Evolution of Tribalism," *Psychological Inquiry* 31, no. 1 (2020): 23–25.
6. "Aerospace Industry": *Final Report of the Commission on the Future of the United States Aerospace Industry*, Commission on the Future of the United States, accessed September 16, 2021, https://www.haydenplanetarium.org/tyson/media/pdf/AeroCommissionFinalReport.pdf.
7. "speak only English": "The Flight of Apollo-Soyuz," NASA, accessed July 20, 2021, https://history.nasa.gov/apollo/apsoyhist.html.
8. "unmarried women": National Center for Health Statistics, "Percent of Babies Born to Unmarried Mothers by State," Centers for Disease Control and Prevention, accessed June 30, 2021, https://www.cdc.gov/nchs/pressroom/sosmap/unmarried/unmarried.htm.
9. "general election": "2000 Presidential Election," accessed June 30, 2021, https://www.270towin.com/2000_Election/.
10. "among the cheaters": Nathan McAlone, "A Chart Made from the Leaked Ashley Madison Data Reveals Which States in the US Like to Cheat the Most," *Insider*, August 20, 2015, accessed June 30, 2021, https://www.businessinsider.com/ashley-madison-leak-reveals-which-states-like-to-cheat-the-most-2015-8.
11. "Republican Platform": *Report of 2018 Permanent Platform & Resolutions Committee*, Republican Party of Texas, accessed December 18, 2021, https://www.texasgop.org/wp-content/uploads/2018/06/PLATFORM-for-voting.pdf.
12. "oil-dependent": More than a third of the Texas economy derives from oil revenues; see Brandon Mulder, "Fact-Check: Is the Texas Oil and Gas Industry 35% of the State Economy?," *Austin American-Statesman*, accessed December 18, 2021, https://www.statesman.com/story/news/politics/politifact/2020/12/22/fact-check-texas-oil-and-gas-industry-35-state-economy/4009134001/.

13. "leaving behind only": *Report of 2020 Platform & Resolutions Committee*, Republican Party of Texas, accessed December 18, 2021, https://drive.google.com/file/d/1HFTbz1vb6KSqwu9Rjv4zxc-85q14XzhZ/view.
14. "climate scientists agree": FYI: this figure has increased to 100 percent of research papers on climate change published since 2019; see James Powell, "Scientists Reach 100% Consensus on Anthropogenic Global Warming," *Bulletin of Science, Technology & Society* 37, no. 4 (November 20, 2019), accessed July 16, 2021, https://journals.sagepub.com/doi/10.1177/0270467619886266.
15. "Green New Deal": Green Party, "Green New Deal," accessed January 1, 2022, https://www.gp.org/green_new_deal.
16. "minority of Christians": Cary Funk, Greg Smith, and David Masci, "How Many Creationists Are There in America?," *Observations* (*Scientific American* blog), February 12, 2019, accessed June 29, 2021, https://blogs.scientificamerican.com/observations/how-many-creationists-are-there-in-america/.
17. "ongoing vaccine programs": Pan American Health Organization, "Measles Elimination in the Americas," accessed June 30, 2021, https://www3.paho.org/hq/index.php?option=com_content&view=article&id=12526:measles-elimination-in-the-americas.
18. "parents refuse": "Measles Resurgence in the United States," Wikipedia, accessed June 30, 2021, https://en.wikipedia.org/wiki/Measles_resurgence_in_the_United_States#Local_outbreaks.
19. "conservative enclaves": Jan Hoffman, "Faith, Freedom, Fear: Rural America's Covid Vaccine Skeptics," *New York Times*, April 30, 2021, accessed June 30, 2021, https://www.nytimes.com/2021/04/30/health/covid-vaccine-hesitancy-white-republican.html.
20. "one-fourth of the nation": Monmouth University Polling Institute, "Public Satisfied with Vaccine Rollout, but 1 in 4 Still Unwilling to Get It," March 8, 2021, accessed June 30, 2021, https://www

.monmouth.edu/polling-institute/reports/monmouthpoll_US _030821/.

21. "vaccine science": *StarTalk Radio*, "Vaccine Science," accessed September 8, 2021, https://www.youtube.com/watch?v=fOOBUixiiac.
22. "accepting audiences": Seth Brown, "Alex Jones's Media Empire Is a Machine Built to Sell Snake-Oil Diet Supplements," *Intelligencer*, May 4, 2017, accessed September 16, 2021, https://nymag.com/intelligencer/2017/05/how-does-alex-jones-make-money.html.
23. "Democratic presidents": Historical Tables, Budget of the United States Government, Fiscal Year 2022, Table 9.8, "Composition of Outlays for the Conduct of Research and Development: 1949–2022," accessed January 2, 2022, https://www.whitehouse.gov/wp-content/uploads/2021/05/hist09z8_fy22.xlsx.
24. "Smithsonian Institution": National Museum of African American History and Culture, "5 Things to Know: HBCU Edition," October 1, 2019, accessed June 2, 2022, https://nmaahc.si.edu/explore/stories/5-things-know-hbcu-edition.
25. "thousands of lynchings": Charles Seguin and David Rigby, "National Crimes: A New National Data Set of Lynchings in the United States, 1883 to 1941," *Socius: Sociological Research for a Dynamic World* 5 (January 1, 2019), accessed September 16, 2021, https://journals.sagepub.com/doi/full/10.1177/2378023119841780.
26. "Clinton and Obama were Black": United States Senate, "Supreme Court Nominations (1789–Present)," United States Senate, accessed April 3, 2022, https://www.senate.gov/legislative/nominations/SupremeCourtNominations1789present.htm.
27. "bluest of blue states": 100 percent of counties in Massachusetts voted blue in the 2020 general election. Politico, "Massachusetts Presidential Results," accessed April 11, 2022, https://www.politico.com/2020-election/results/massachusetts/.
28. "federal government": "Federal Spending by State 2022,"

World Population Review, accessed January 3, 2022, https://worldpopulationreview.com/state-rankings/federal-spending-by-state.

29. "red states": In the 2020 general election.
30. "300,000 people": Rob Salkowitz, "Fans Turn Up for New York Comic Con Even if Big Names Don't," *Forbes*, October 9, 2021, accessed December 17, 2021, https://www.forbes.com/sites/robsalkowitz/2021/10/09/fans-turn-up-for-new-york-comic-con-even-if-big-names-dont.
31. "millions": "Convention Schedule," FanCons, accessed December 17, 2021, https://fancons.com/events/schedule.php?type=all&year=2022&loc=eu.

CHAPTER FIVE: RISK & REWARD

1. "take an average": Thomas Simpson, "A Letter [. . .] on the Advantage of Taking the Mean of a Number of Observations, in Practical Astronomy," *Philosophical Transactions (1683–1775)* 49 (1755–1756), 82–93.
2. "rational evolutionary roots": Michael Shermer, *Conspiracy: Why the Rational Believe the Irrational* (Baltimore: Johns Hopkins University Press, 2022).
3. "4,000 physicists": Stephen Skolnick, "How 4,000 Physicists Gave a Vegas Casino Its Worst Week Ever," *Physics Buzz* (blog), September 10, 2015, accessed June 5, 2022, http://physicsbuzz.physicscentral.com/2015/09/one-winning-move.html.
4. "casino on property": Steve Beauregard, "Biggest Casino in Las Vegas & List of the Top 20 Largest Casinos in Sin City," Gamboool, accessed July 15, 2021, https://gamboool.com/biggest-casinos-in-las-vegas-list-of-the-top-20-largest-casinos-in-sin-city.
5. "$45 billion": Will Yakowicz, "U.S. Gambling Revenue to Break $44 Billion Record in 2021," *Forbes*, August 10, 2021, accessed January 1, 2022, https://www.forbes.com/sites/willyakowicz/2021/08/10/us-gambling-revenue-to—break-44-billion-record-in-2021.

6. "lottery": "Lotteries in the United States," Wikipedia, accessed July 15, 2021, https://en.wikipedia.org/wiki/Lotteries_in_the_United_States#States_with_no_lotteries.
7. "1 in 292.2 million": Investopedia, "The Lottery: Is It Ever Worth Playing?," accessed July 15, 2021, https://www.investopedia.com/managing-wealth/worth-playing-lottery/.
8. "Recent surveys": Erin Richards, "Math Scores Stink in America. Other Countries Teach It Differently—and See Higher Achievement," *USA Today*, February 28, 2020, accessed January 5, 2022, https://www.usatoday.com/story/news/education/2020/02/28/math-scores-high-school-lessons-freakonomics-pisa-algebra-geometry/4835742002/.
9. "inaugural Tweet": Neil deGrasse Tyson (@neiltyson), Twitter, February 9, 2010, 3:46 p.m., accessed July 15, 2021, https://twitter.com/neiltyson/status/8870114781.
10. "major averages rose": CNBC (@CNBC), Twitter, December 10, 2021, 4:03 p.m., accessed May 17, 2022, https://twitter.com/CNBC/status/1469412512357568521.
11. "many more than that": TipRanks, "2 'Strong Buy' Stocks from a Top Wall Street Analyst," July 13, 2021, accessed July 15, 2021, https://www.yahoo.com/now/2-strong-buy-stocks-top-091615572.html.
12. "sites that rank traders": TipRanks, "Top Wall Street Analysts," accessed July 23, 2021, https://www.tipranks.com/analysts/top.
13. "more than 30 percent": Sam Ro, "The Truth About Warren Buffett's Investment Track Record," Yahoo! Finance, March 1, 2021, accessed December 21, 2021, https://www.yahoo.com/now/the-truth-about-warren-buffetts-investment-track-record-morning-brief-113829049.html.
14. "Scientists and right-leaning people": National Academies of Sciences, Engineering, and Medicine, *Genetically Engineered Crops: Experiences and Prospects* (Washington, DC: National Academies Press, 2016), accessed July 15, 2021, https://www

.nationalacademies.org/our-work/genetically-engineered-crops-past-experience-and-future-prospects.

15. "narrate a documentary": *Food Evolution*, directed by Scott Hamilton Kennedy, narrated by Neil deGrasse Tyson (Black Valley Films, 2016).
16. "GMO corn syrup": "Ben & Jerry's Statement on Glyphosate," Ben & Jerry's (website), accessed July 15, 2021, https://www.benjerry.com/about-us/media-center/glyphosate-statement.
17. "salty ocean water": Samuel Taylor Coleridge, *Rime of the Ancient Mariner*, part 2, stanza 9 (1817).
18. "pains to human nature": Walter Bagehot, *Physics and Politics,* No. V: "The Age of Discussion" (Westport, CT: Greenwood Press, 1872).
19. "American Cancer Society's web page": American Cancer Society, "Colorectal Cancer Risk Factors," accessed July 16, 2021, https://www.cancer.org/cancer/colon-rectal-cancer/causes-risks-prevention/risk-factors.html.
20. "colorectal cancer": American Cancer Society, "Key Statistics for Colorectal Cancer," accessed July 16, 2021, https://www.cancer.org/cancer/colon-rectal-cancer/about/key-statistics.html.
21. "meta-study of research": Manuela Chiavarini et al., "Dietary Intake of Meat Cooking-Related Mutagens (HCAs) and Risk of Colorectal Adenoma and Cancer: A Systematic Review and Meta-Analysis," *Nutrients* 9, no. 5 (May 18, 2017): 515, accessed June 7, 2022, https://www.ncbi.nlm.nih.gov/pmc/articles/PMC5452244/.
22. "1-in-8 likelihood": Hannah Ritchie and Max Roser, "Smoking," *Our World in Data*, May 2013, revised November 2019, accessed June 30, 2021, https://ourworldindata.org/smoking; see also Lynne Eldridge, "What Percentage of Smokers Get Lung Cancer?," Verywell Health, accessed June 28, 2021, https://www.verywellhealth.com/what-percentage-of-smokers-get-lung-cancer-2248868.

23. "the suburbs": John Woodrow Cox and Steven Rich, "Scarred by School Shootings," *Washington Post*, updated March 25, 2018, accessed July 15, 2021, https://www.washingtonpost.com/graphics/2018/local/us-school-shootings-history/.
24. "safer in the city": William H. Lucy, "Mortality Risk Associated with Leaving Home: Recognizing the Relevance of the Built Environment," *American Journal of Public Health* 93, no. 9 (September 2003): 1564–69, accessed July 16, 2021, https://www.ncbi.nlm.nih.gov/pmc/articles/PMC1448011/; Bryan Walsh, "In Town vs. Country, It Turns Out That Cities Are the Safest Places to Live," *Time*, July 23, 2013, accessed July 16, 2021, https://science.time.com/2013/07/23/in-town-versus-country-it-turns-out-that-cities-are-the-safest-places-to-live/.
25. "22 percent higher": Sage R. Meyers et al., "Safety in Numbers: Are Major Cities the Safest Places in the United States?" *Injury Prevention* 62, no. 4 (October 1, 2013): 408–18.E3, accessed June 8, 2022, https://www.ncbi.nlm.nih.gov/pmc/articles/PMC3993997/.
26. "Texas, shooting": "2019 El Paso Shooting," Wikipedia, accessed July 16, 2021, https://en.wikipedia.org/wiki/2019_El_Paso_shooting.
27. "war on terror": Paulina Cachero, "US Taxpayers Have Reportedly Paid an Average of $8,000 Each and over $2 Trillion Total for the Iraq War Alone," *Insider*, February 6, 2020, accessed July 16, 2021, https://www.businessinsider.com/us-taxpayers-spent-8000-each-2-trillion-iraq-war-study-2020-2.
28. "white-tailed deer": Sophie L. Gilbert et al., "Socioeconomic Benefits of Large Carnivore Recolonization Through Reduced Wildlife-Vehicle Collisions," *Conservation Letters* 10, no. 4 (July/August 2017): 431–39, accessed August 1, 2021, https://conbio.onlinelibrary.wiley.com/doi/epdf/10.1111/conl.12280.
29. "crashes around the world": Murat Karacasu and Arzu Er, "An Analysis on Distribution of Traffic Faults in Accidents, Based on

Driver's Age and Gender: Eskisehir Case" *Procedia—Social and Behavioral Sciences 20* (2011), 776–785, accessed June 8, 2022, https://www.sciencedirect.com/science/article/pii/S1877042811014662.

30. "Tesla Says Autopilot": Neal E. Boudette, "Tesla Says Autopilot Makes Its Cars Safer; Crash Victims Say It Kills," *New York Times*, July 5, 2021, accessed July 27, 2021, https://www.nytimes.com/2021/07/05/business/tesla-autopilot-lawsuits-safety.html.
31. "more than 1,000": "List of Fatal Accidents and Incidents Involving Commercial Aircraft in the United States," Wikipedia, accessed July 27, 2021, https://en.wikipedia.org/wiki/List_of_fatal_accidents_and_incidents_involving_commercial_aircraft_in_the_United_States.
32. "without a single crash": Highly publicized crashes of the Boeing 737 MAX in 2018 and 2019 involved non-US-based carriers, and so don't contribute to this statistic.
33. "died from other causes": Leslie Josephs, "The Last Fatal US Airline Crash Was a Decade Ago; Here's Why Our Skies Are Safer," CNBC, February 13, 2019, updated March 8, 2019, accessed July 27, 2021, https://www.cnbc.com/2019/02/13/colgan-air-crash-10-years-ago-reshaped-us-aviation-safety.html.
34. "risen 35 percent from 2000": Bureau of Transportation Statistics, United States Department of Transportation, "U.S. Air Carrier Traffic Statistics Through November 2021," accessed August 1, 2021, https://www.transtats.bts.gov/TRAFFIC/.
35. "both sides now": Joni Mitchell, stanza from the song "Both Sides Now" (Detroit: Gandalf Publishing, 1967).

CHAPTER SIX: MEATARIANS & VEGETARIANS

1. "*Beef* magazine": Paul Copan, Wes Jamison, and Walter Kaiser, *What Would Jesus Really Eat: The Biblical Case for Eating Meat* (Burlington, ON: Castle Quay Books, 2019); see also Amanda Radke, "Yes, Jesus Would Eat Meat & You Can, Too," *Beef*, June 9, 2022, accessed April 14, 2022, https://

www.beefmagazine.com/beef/yes-jesus-would-eat-meat-you-can-too.

2. "vegetarians in the world": "Vegetarianism by Country," Wikipedia, accessed August 7, 2021, https://en.wikipedia.org/wiki/Vegetarianism_by_country.
3. "3 percent of the population": RJ Reinhart, "Snapshot: Few Americans Vegetarian or Vegan," Gallup, August 1, 2018, accessed August 7, 2021, https://news.gallup.com/poll/238328/snapshot-few-americans-vegetarian-vegan.aspx.
4. "meat consumption has tripled": Hannah Ritchie and Max Roser, "Meat and Dairy Production," *Our World in Data*, August 2017, revised November 2019, accessed August 11, 2021, https://ourworldindata.org/meat-production.
5. "brought to slaughter": "What Is the Age Range for Butchering Steers? I Am Trying for Prime," Beef Cattle, September 3, 2019, accessed August 7, 2021, https://beef-cattle.extension.org/what-is-the-age-range-for-butchering-steers-i-am-trying-for-prime.
6. "acres of land": University of California Cooperative Extension, "Sample Costs for a Cow-Calf/Grass-Fed Beef Operation," 2004, accessed February 25, 2022, https://coststudyfiles.ucdavis.edu/uploads/cs_public/83/84/838417e7-bdad-40e6-bcaa-c3d80ccdcd71/beefgfnc2004.pdf.
7. "cattle into 800 acres": "The Biggest CAFO in the United State," Wickersham's Conscience, March 20, 2020, accessed April 14, 2022, https://wickershamsconscience.wordpress.com/2020/03/20/the-biggest-cafo-in-the-united-states/.
8. "500 pounds of meat": South Dakota State University Extension, "How Much Meat Can You Expect from a Fed Steer?," updated August 6, 2020, accessed June 9, 2022, https://extension.sdstate.edu/how-much-meat-can-you-expect-fed-steer.
9. "reply are elsewhere": Neil deGrasse Tyson, *Letters from an Astrophysicist* (New York: W. W. Norton, 2019).

10. "And God said": Genesis 1:26, King James Version.
11. "With rare exceptions": See, e.g., Ryan Patrick McLaughlin, "A Meatless Dominion: Genesis 1 and the Ideal of Vegetarianism," *Biblical Theology Bulletin* 47, no. 3 (August 2, 2017): 144–54, accessed August 7, 2021, https://journals.sagepub.com/doi/10.1177/0146107917715587.
12. "stewardship": Eric O'Grey, "Vegan Theology for Christians," PETA Prime, January 30, 2018, accessed August 7, 2021, https://prime.peta.org/2018/01/vegan-theology-christians/.
13. "subfield of academic philosophy": Peter Singer, *Animal Liberation* (New York: Harper Collins, 1975).
14. "subject of persistent activism": PETA (website), accessed August 7, 2021, https://www.peta.org.
15. "snails one evening": Story recounted live on *StarTalk*, August 22, 2011, https://www.startalkradio.net/show/making-the-fur-fly/.
16. "light, which attracts them": "Do Snails Have Eyes?," Facts About Snails, accessed August 9, 2021, https://factsaboutsnails.com/snail-facts/do-snails-have-eyes/.
17. "caught in the net": Rene Ebersole, "How 'Dolphin Safe' Is Canned Tuna, Really?," *National Geographic*, March 10, 2021, accessed August 8, 2021, https://www.nationalgeographic.com/animals/article/how-dolphin-safe-is-canned-tuna.
18. "safety of your home": Animal Diversity Web, University of Michigan, Museum of Zoology, "*Mus musculus* house mouse," accessed April 24, 2022, https://animaldiversity.org/accounts/Mus_musculus/.
19. "fifty full-grown trees": "Learn How Many Trees It Takes to Build a House?," Home Preservation Manual, accessed August 9, 2021, https://www.homepreservationmanual.com/how-many-trees-to-build-a-house/.
20. "lived a half century": Michael H. Ramage et al., "The Wood from the Trees: The Use of Timber in Construction," *Renewable and Sustainable Energy Reviews* 68 (February 2017): 333,

accessed January 20, 2022, https://www.sciencedirect.com/science/article/pii/S1364032116306050.

21. "oxygen-producing plant life": Kyle Cunningham, "Landowner's Guide to Determining Weight of Standing Hardwood Trees," University of Arkansas Division of Agriculture, Cooperative Extension Service, accessed August 9, 2021, https://www.uaex.uada.edu/publications/pdf/FSA-5021.pdf.
22. "concentrated maple tree blood": "Maple Syrup Concentration," Synder Filtration, accessed January 30, 2021, https://synderfiltration.com/2014/wp-content/uploads/2014/07/Maple-Syrup-Concentration-Case-Study.pdf.
23. "root system of the forest": Britt Holewinski, "Underground Networking: The Amazing Connections Beneath Your Feet," National Forest Foundation, accessed January 30, 2022, https://www.nationalforests.org/blog/underground-mycorrhizal-network.
24. "good authority": Steven Spielberg, private communication, April 2004, Hayden Planetarium, New York City.
25. "quality children's television": Associated Press, "Lewis Throws Voice to Push for Quality TV," *Deseret News*, March 11, 1993, accessed September 8, 2021, https://www.deseret.com/1993/3/11/19036574/lewis-throws-voice-to-push-for-quality-tv.
26. "publicly traded": Mitch Zinck, "Top 10 Stocks to Invest in Lab-Grown Meat," Lab Grown Meat, June 29, 2021, accessed August 11, 2021, https://labgrownmeat.com/top-10-stocks/.
27. "contains the following attack": Chuck Lorre, "Card #536," Chuck Lorre Productions, Official Vanity Card Archives, September 26, 2016, accessed September 16, 2021, http://chucklorre.com/?e=980.
28. "I suppose no body": Christiaan Huygens, *The Celestial Worlds Discover'd: or, Conjectures Concerning the Inhabitants, Plants and Productions of the Worlds in the Planets* (London: Timothy Childe, 1698), accessed June 10, 2022, https://galileo.ou.edu

/exhibits/celestial-worlds-discoverd-or-conjectures-concerning-inhabitants-plants-and-productions.

29. "dialogue captures the astonishment": Terry Bisson, *They're Made out of Meat, and 5 Other All-Talk Tales* (Amazon.com, Kindle edition, 2019).

CHAPTER SEVEN: GENDER & IDENTITY

1. "Schrödinger's cat": "Schrödinger's Cat," Wikipedia, accessed July 6, 2022, https://en.wikipedia.org/wiki/Schrödinger's_cat.
2. "nonconforming designations": "What Does LGBTQ+ Mean?," OK2BME, accessed August 22, 2021, https://ok2bme.ca/resources/kids-teens/what-does-lgbtq-mean/.
3. "*West Side Story*": First produced for Broadway in 1957.
4. "Bible has a verse about it": Deuteronomy 22:5, King James Version.
5. "burned at the stake": "Trial of Joan of Arc," Wikipedia, accessed April 24, 2022, https://en.wikipedia.org/wiki/Trial_of_Joan_of_Arc.
6. "rest of the animal kingdom": Joan Roughgarden, *Evolution's Rainbow: Diversity, Gender, and Sexuality in Nature and People*, 10th anniversary ed. (Berkeley: University of California Press, 2013).
7. "jaw and brow relative to women": Anthony C. Little, Benedict C. Jones, and Lisa M. DeBruine, "Facial Attractiveness: Evolutionary Based Research," *Philosophical Transactions of the Royal Society B: Biological Sciences* 366, no. 1571 (June 12, 2011): 1638–59, accessed March 18, 2022, https://www.ncbi.nlm.nih.gov/pmc/articles/PMC3130383/.
8. "women do annually": American Society of Plastic Surgeons, *Plastic Surgery Statistics Report*, 2020, accessed November 28, 2021, https://www.plasticsurgery.org/documents/News/Statistics/2020/plastic-surgery-statistics-full-report-2020.pdf.

9. "well before Christmas": US Food and Drug Administration, "Fun Facts About Reindeer and Caribou," content current as of February 13, 2020, accessed December 21, 2021, https://www.fda.gov/animal-veterinary/animal-health-literacy/fun-facts-about-reindeer-and-caribou.
10. "some of which are feminine": "What Are the Names of Santa's Reindeer?," Iglu Ski, accessed December 21, 2021, https://www.igluski.com/lapland-holidays/what-are-the-names-of-santas-reindeer.
11. "five categories": Saffir-Simpson Hurricane Wind Scale, National Hurricane Center and Central Pacific Hurricane Center, accessed January 5, 2022, https://www.nhc.noaa.gov/aboutsshws.php.
12. "sport even more colors": Ariane Resnick, "What Do the Colors of the New Pride Flag Mean?," Verywell Mind, updated June 21, 2021, accessed August 22, 2021, https://www.verywellmind.com/what-the-colors-of-the-new-pride-flag-mean-5189173.
13. "expression of their gender identity": Tom Dart, "Texas Clings to Unconstitutional Homophobic Laws—and It's Not Alone," *Guardian*, June 1, 2019, accessed September 17, 2021, https://www.theguardian.com/world/2019/jun/01/texas-homophobic-laws-lgbt-unconstitutional.

CHAPTER EIGHT: COLOR & RACE

1. "bookkeeping of all the data": For a full history of this period, see Dava Sobel, *The Glass Universe: How the Ladies of the Harvard Observatory Took the Measure of the Stars* (New York: Viking, 2016).
2. "high in the stratosphere": Jennifer Chu, "Study: Reflecting Sunlight to Cool the Planet Will Cause Other Global Changes," *MIT News*, June 2, 2020, accessed August 20, 2021, https://news.mit.edu/2020/reflecting-sunlight-cool-planet-storm-0602.
3. "latitude on Earth and skin tone": Nina Jablonski and George Chaplin, "The Colours of Humanity: The Evolution of

Pigmentation in the Human Lineage," *Philosophical Transactions of the Royal Society B* 372 (May 22, 2017), accessed August 22, 2021, https://royalsocietypublishing.org/doi/pdf/10.1098/rstb.2016.0349.

4. "Earth's surface at that latitude": Nina Jablonski and George Chaplin, "Human Skin Pigmentation as an Adaptation to UV Radiation," *Proceedings of the National Academy of Sciences* 107, Suppl. 2 (May 5, 2010), accessed August 22, 2021, https://doi.org/10.1073/pnas.0914628107.
5. "genomic pathways": Nicholas G. Crawford et al., "Loci Associated with Skin Pigmentation Identified in African Populations," *Science* 358, no. 6365 (October 12, 2017), accessed January 31, 2022, https://www.science.org/doi/10.1126/science.aan8433.
6. "hair color within a brand": Clairol, "Natural Instincts" semi-permanent hair color.
7. "thousands of hues": Benjamin Moore (website), accessed July 6, 2022, https://www.benjaminmoore.com/en-us/color-overview/find-your-color/color-families.
8. "the South's population": "1860 United States Census," Wikipedia, accessed September 24, 2021, https://en.wikipedia.org/wiki/1860_United_States_census.
9. "The greatest of all": James Henry Hammond, "On the Question of Receiving Petitions on the Abolition of Slavery in the District of Columbia," Address to Congress, February 1, 1836, accessed March 19, 2022, https://babel.hathitrust.org/cgi/pt?id=hvd.hx4q2m.
10. "American Museum of Natural History": Where I've served as the Frederick P. Rose Director of the Hayden Planetarium since 1996.
11. "The problem is": Theodore Roosevelt, "Lincoln and the Race Problem," speech to the New York Republican Club, February 13, 1905, accessed September 8, 2021, https://www.blackpast.org/african-american-history/1905-theodore-roosevelt-lincoln-and-race-problem-3/.
12. "America for decades": American Museum of Natural History,

"Museum Statement on Eugenics," September 2021, accessed July 6, 2022, https://www.amnh.org/about/eugenics-statement.

13. "bench along Bond Street": "Allies Sculpture," Atlas Obscura, accessed September 8, 2021, https://www.atlasobscura.com/places/allies.
14. "embarrassingly patronizing": In 2020, a long-standing replica of *Emancipation* was removed from Park Square in Boston, Massachusetts, to the sculptor's hometown after protests.
15. "The two figures": American Museum of Natural History, "What Did the Artists and Planners Intend?," accessed July 6, 2022, https://www.amnh.org/exhibitions/addressing-the-statue/artist-intent.
16. "new presidential library": In-house memo to museum staff, accessed July 6, 2022, https://en.wikipedia.org/wiki/Theodore_Rosevelt_Presidential_Library.
17. "feelings of superiority": Meilan Solly, "DNA Pioneer James Watson Loses Honorary Titles over Racist Comments," *Smithsonian*, January 15, 2019, accessed September 19, 2021, https://www.smithsonianmag.com/smart-news/dna-pioneer-james-watson-loses-honorary-titles-over-racist-comments-180971266/.
18. "hairy ball": "Hairy Ball Theorem," Wikipedia, accessed September 8, 2021, https://en.wikipedia.org/wiki/Hairy_ball_theorem.
19. "exclusive entry on just this topic": "List of Electronic Color Code Mnemonics," Wikipedia, accessed January 5, 2022, https://en.wikipedia.org/wiki/List_of_electronic_color_code_mnemonics.
20. "Worth of Different Races": Francis Galton, *Hereditary Genius: An Inquiry into Its Laws and Consequences* (New York: D. Appleton, 1870), 339.
21. "90 percent were enslaved": Aaron O'Neill, "Black and Slave Population of the United States from 1790 to 1880," Statista, March 19, 2021, accessed September 12, 2021, https://www.statista.com/statistics/1010169/black-and-slave-population-us-1790–1880/.

22. "Comparing them by their faculties": Thomas Jefferson, *Notes on the State of Virginia* (Baltimore: W. Pechin, 1800), 151.
23. "producing six children": Monticello, "The Life of Sally Hemings," accessed September 12, 2021, https://www.monticello.org/sallyhemings/.
24. "*The Origin of Races*": Carleton S. Coon, *The Origin of Races* (New York: Alfred A. Knopf, 1962), 656.
25. "ascending their backs": "Men with Hairy Chest," DC Urban Moms and Dads, December 23, 2014, accessed September 12, 2021, https://www.dcurbanmom.com/jforum/posts/list/435718.page.
26. "shade of black or brown": Toshisada Nishida, "Chimpanzee," *Encyclopedia Britannica*, accessed September 12, 2021, https://www.britannica.com/animal/chimpanzee.
27. "a fraction of 1 percent": Medline Plus, "What Does It Mean to Have Neanderthal or Denisovan DNA?," accessed September 12, 2021, https://medlineplus.gov/genetics/understanding/dtcgenetictesting/neanderthaldna/. Also Chen, Lu, et al., "Identifying and Interpreting Apparent Neanderthal Ancestry in African Individuals", *Cell* 180, no. 4 (February 20, 2020): 677–87, accessed July 12, 2023, https://www.cell.com/cell/fulltext/S0092-8674(20)30059-3.
28. "technologies to shape their world": Angela Saini, *Superior: The Return of Race Science* (Boston: Beacon Press, 2019), 18–20.
29. "Probably not": "Can African Americans Get Head Lice?," Lice Aunties, April 14, 2021, accessed September 12, 2021, https://liceaunties.com/can-african-americans-get-head-lice/; see also W. Wayne Price and Amparo Benitez, "Infestation and Epidemiology of Head Lice in Elementary Schools in Hillsborough Country, Florida," *Florida Scientist* 52, no. 4 (1989): 278–88.
30. "hair of Black people": Robin A. Weiss, "Apes, Lice and Prehistory," *Journal of Biology* 8, no. 20 (2009), accessed September 21, 2021, https://www.ncbi.nlm.nih.gov/pmc/articles/PMC2687769/.

31. "contract skin cancer": United States Cancer Statistics, "Leading Cancers by Age, Sex, Race and Ethnicity 2019," Centers for Disease Control and Prevention, accessed July 6, 2022, https://gis.cdc.gov/Cancer/USCS/#/Demographics/; see also: Healthline, "Yes, Black People Can Get Skin Cancer. Here's What to Look For," accessed April 11, 2022, https://www.healthline.com/health/skin-cancer/can-black-people-get-skin-cancer.
32. "White people as among Black people": Joel M. Gelfand et al., "The Prevalence of Psoriasis in African Americans: Results from a Population-Based Study," *Journal of the American Academy of Dermatology* 52, no. 1 (2005): 23, accessed September 18, 2021, https://pubmed.ncbi.nlm.nih.gov/15627076/.
33. "onset of osteoporosis": Bone Health and Osteoporosis Foundation, "What Is Osteoporosis and What Causes It?," accessed July 6, 2022, https://www.bonehealthandosteoporosis.org/patients/what-is-osteoporosis/.
34. "bones remain good and strong": J. F. Aloia et al., "Risk for Osteoporosis in Black Women," *Calcified Tissue International* 59 (1996): 415–23, accessed September 18, 2021, https://link.springer.com/article/10.1007%2FBF00369203.
35. "suicide rate of Black teens": Jeffrey A. Bridge et al., "Suicide Trends Among Elementary School-Aged Children in the United States from 1993 to 2012," *JAMA Pediatrics* 169, no. 7 (2015): 673–77, accessed July 6, 2022, https://jamanetwork.com/journals/jamapediatrics/fullarticle/2293169.
36. "higher than for Black people": Suicide Prevention Resource Center, "Racial and Ethnic Disparities," accessed September 18, 2021, https://sprc.org/scope/racial-ethnic-disparities.
37. "anorexic eating disorders": Many studies, for example: Jacquelyn Y. Taylor et al., "Prevalence of Eating Disorders Among Blacks in the National Survey of American Life," *International Journal of Eating Disorders* 40 (2007 Suppl.): S10–S14, accessed July 6, 2022, https://www.ncbi.nlm.nih.gov/pmc/articles/PMC2882704;

see also Ruth H. Striegel-Moore et al., "Eating Disorders in White and Black Women," *American Journal of Psychiatry* 160 (2003): 1326–31, accessed July 6, 2022, https://pubmed.ncbi.nlm.nih.gov/12832249/.

38. "origins story, which I paraphrase": Keb Meh, "Mythologies of Skin Color and Race in Ethiopia," *Japan Sociology*, December 2, 2014, accessed September 12, 2021, https://japansociology.com/2014/12/02/mythologies-of-skin-color-and-race-in-ethiopia/.
39. "just over four feet tall": Guinness World Records, "Shortest Tribe," accessed September 12, 2021, https://www.guinnessworldrecords.com/world-records/shortest-tribe.
40. "stands six feet tall": Guinness World Records, "Tallest Tribe," accessed September 12, 2021, https://www.guinnessworldrecords.com/world-records/67503-tallest-tribe.
41. "taller than that of Indonesians": "Average Human Height by Country," Wikipedia, accessed July 6, 2022, https://en.wikipedia.org/wiki/Average_human_height_by_country.
42. "denial among White people": Ben McGrath, "Did Spacemen, or People with Ramps, Build the Pyramids?," *New Yorker*, August 23, 2021, accessed February 26, 2022, https://www.newyorker.com/magazine/2021/08/30/did-spacemen-or-people-with-ramps-build-the-pyramids.
43. "an African himself": Elon Musk is from South Africa, a place where native-born White people hardly ever refer to themselves as Africans, but of course they all are. Elon Musk (@elonmusk), Twitter, July 31, 2020, 12:14 a.m., accessed April 11, 2022, https://twitter.com/elonmusk/status/1289051795763769345?lang=en.
44. "African Institute for Mathematical Sciences": https://nexteinstein.org; see also "Neil Turok Bets the Next Einstein Will Be from Africa," TED Prize-Winning Wishes, 2008, accessed September 12, 2021, https://www.ted.com/participate/ted-prize/prize-winning-wishes/aims-next-einstein-initiative.

45. "African Einstein": Neil Turok, "Africa AIMS High," *Nature* 474 (2011): 567, accessed September 12, 2021, https://www.nature.com/articles/474567a.
46. "United Arab Emirates": International Chess Federation, "Top Chess Federations," accessed December 28, 2021, https://ratings.fide.com/top_federations.phtml.
47. "Zambia $1,000": World Bank, "GDP per Capita," accessed December 28, 2021, https://data.worldbank.org/indicator/NY.GDP.PCAP.CD.
48. "among the top 4 percent": International Chess Federation, "Rating Analytics: The Number of Rated Chess Players Goes Up," accessed December 28, 2021, https://www.fide.com/news/288; see also "FIDE Titles," Wikipedia, accessed December 28, 2021, https://en.wikipedia.org/wiki/FIDE_titles.
49. "higher household income": Molly Fosco, "The Most Successful Ethnic Group in the U.S. May Surprise You," IMDiversity, June 7, 2018, accessed April 7, 2022, https://imdiversity.com/diversity-news/the-most-successful-ethnic-group-in-the-u-s-may-surprise-you/.
50. "White UK counterparts": Jill Rutter, "Back to Basics: Towards a Successful and Cost-Effective Integration Policy," Report: Institute for Public Policy Research, UK, March 2013; see also "GCSE English and Math's Results March 2022," Department of Education, UK, accessed April 7, 2022, https://www.ethnicity-facts-figures.service.gov.uk/education-skills-and-training/11-to-16-years-old/a-to-c-in-english-and-maths-gcse-attainment-for-children-aged-14-to-16-key-stage-4/latest.
51. "people existed in the world": Statista, "Estimated Global Population from 10,000 BCE to 2100," accessed September 12, 2021, https://www.statista.com/statistics/1006502/global-population-ten-thousand-bc-to-2050/.
52. "New York metropolitan area": Statista, "Population of the New York-Newark-Jersey City Metro Area in the United States from

2010 to 2020," accessed January 4, 2022, https://www.statista.com/statistics/815095/new-york-metro-area-population/.

53. "no descendants at all": Dr. Yan Wong, "Family Trees: Tracing the World's Ancestor," BBC, August 22, 2012, accessed September 12, 2021, https://www.bbc.com/news/magazine-19331938.
54. "I Have a Dream": "Read Martin Luther King Jr.'s 'I Have a Dream' Speech in Its Entirety," NPR, accessed January 4, 2022, https://www.npr.org/2010/01/18/122701268/i-have-a-dream-speech-in-its-entirety.
55. "I do not pretend": Reverend Theodore Parker, "Of Justice and the Conscience," in *Ten Sermons of Religion* (Boston: Crosby, Nichols, 1853), 85, accessed July 6, 2022, http://www.fusw.org/uploads/1/3/0/4/13041662/of-justice-and-the-conscience.pdf; see also *All Things Considered*, "Theodore Parker and the Moral Universe," NPR, September 2, 2010, accessed September 8, 2021, https://www.npr.org/templates/story/story.php?storyId=129609461.

CHAPTER NINE: LAW & ORDER

1. "trial by water": "The Code of Hammurabi," translated by L. W. King, Avalon Project, Yale Law School, accessed December 20, 2021, https://avalon.law.yale.edu/ancient/hamframe.asp.
2. "Antonio Pigafetta's eyewitness account": Antonio Pigafetta, "Navigation," in *Magellan's Voyage: A Narrative Account of the First Circumnavigation*, translated and edited by R. A. Skelton (1519; New York: Dover, 1969), 147.
3. "The great tragedy": "Address to the British Association for the Advancement of Science," delivered by the president, Thomas H. Huxley (Liverpool, September 15, 1870), accessed December 21, 2021, http://aleph0.clarku.edu/huxley/CE8/B-Ab.html.
4. "Blackstone ratio declares": William Blackstone, *Commentaries on the Laws of England* (Oxford: Clarendon Press, 1765).

5. "lawful judgment of his peers": "Magna Carta: Muse and Mentor (Trial by Jury)," Library of Congress exhibition, 2014–15, accessed December 20, 2021, https://www.loc.gov/exhibits/magna-carta-muse-and-mentor/trial-by-jury.html; see also, *The Online Library of Liberty*, accessed July 6, 2022, https://oll-resources.s3.us-east-2.amazonaws.com/oll3/store/titles/2142/Blackstone_1387-02_EBk_v6.0.pdf.
6. "eyewitness testimony": For example: Matthew J. Sharps, "Eyewitness Testimony, Eyewitness Mistakes: What We Get Wrong," *Psychology Today*, August 21, 2020, accessed December 14, 2021, https://www.psychologytoday.com/us/blog/the-forensic-view/202008/eyewitness-testimony-eyewitness-mistakes-what-we-get-wrong.
7. "The Innocence Project's mission": Innocence Project (website), accessed July 6, 2022, https://innocenceproject.org/exonerate/.
8. "executed over that time": Equal Justice Initiative, "Death Penalty," accessed January 2, 2022, https://eji.org/issues/death-penalty/.
9. "behind bars": Innocence Project, "DNA Exonerations in the United States," accessed December 20, 2021, https://innocenceproject.org/dna-exonerations-in-the-united-states/.
10. "faulty forensic science": National Research Council, *Strengthening Forensic Science in the United States: A Path Forward* (Washington, DC: National Academies Press, 2009), accessed December 20, 2021, https://doi.org/10.17226/12589.
11. "misidentifying him": Alison Flood, "Alice Sebold Publisher Pulls Memoir After Overturned Rape Conviction," *Guardian*, December 1, 2021, accessed December 20, 2021, https://www.theguardian.com/books/2021/dec/01/alice-sebold-publisher-pulls-memoir-overturned-conviction-lucky-anthony-broadwater.
12. "prisoners in the US": Ann E. Carson, "Prisoners in 2020," US Department of Justice, Bureau of Justice Statistics, December

2021, accessed March 6, 2022, https://bjs.ojp.gov/content/pub/pdf/p20st.pdf.

13. "worldwide": Roy Walmsley and Helen Fair, "World Prison Population List," 13th Edition, December 2021, accessed March 6, 2022, https://www.prisonstudies.org/sites/default/files/resources/downloads/world_prison_population_list_13th_edition.pdf; Roy Walmsley, "World Female Imprisonment List," 4th Edition, November 2017, Institute for Criminal Policy Research, UK, accessed March 6, 2022, https://www.prisonstudies.org/sites/default/files/resources/downloads/world_female_prison_4th_edn_v4_web.pdf.
14. "Y chromosome": Medline Plus, "Y Chromosome," accessed March 6, 2022, https://medlineplus.gov/genetics/chromosome/y.
15. "Starmus science festival": Starmus (website), accessed July 2, 2021, https://www.starmus.com.
16. "headlines": Robert F. Graboyes, "The Rationalia Fallacy," *U.S. News & World Report*, July 18, 2016, accessed June 30, 2021, https://www.usnews.com/opinion/articles/2016-07-18/neil-degrasse-tyson-may-dream-of-a-rationalia-society-but-its-a-fallacy; Jeffrey Guhin, "A Nation Ruled by Science Is a Terrible Idea," *Slate*, July 5, 2016, accessed June 30, 2021, https://slate.com/technology/2016/07/neil-degrasse-tyson-wants-a-nation-ruled-by-evidence-but-evidence-explains-why-thats-a-terrible-idea.html; G. Shane Morris, "Neil DeGrasse Tyson's 'Rationalia' Would Be a Terrible Country," *Federalist*, July 1, 2016, accessed June 30, 2021, https://thefederalist.com/2016/07/01/neil-degrasse-tysons-rationalia-would-be-a-terrible-country/; "Sorry, Neil deGrasse Tyson, Basing a Country's Governance Solely on 'The Weight of Evidence' Could Not Work," ArtsJournal, June 30, 2016, accessed June 30, 2021, https://www.artsjournal.com/2016/07/sorry-neil-degrasse-tyson-basing-a-countrys-governance-solely-on-the-weight-of-evidence-could-not-work.html.

CHAPTER TEN: BODY & MIND

1. "*Physician's Desk Reference*": Now available digitally, https://www.pdr.net.
2. "your beating heart": Medical News Today, "How Long You Can Live Without Water," accessed November 29, 2021, https://www.medicalnewstoday.com/articles/325174.
3. "ligaments and tendons": UpToDate, "Bones of the Foot," accessed December 12, 2021, https://www.uptodate.com/contents/image?imageKey=SM%2F52540&topicKey=SM%2F17003.
4. "cells of our own bodies": Zhi Y. Kho and Sunil K. Lal, "The Human Gut Microbiome—A Potential Controller of Wellness and Disease," *Frontiers of Microbiology* 9 (2018), accessed November 28, 2021, https://doi.org/10.3389/fmicb.2018.01835.
5. "pass into your bloodstream": Associated Press, "Chocolate Cravings May Be a Real Gut Feeling," NBC News, October 12, 2007, accessed November 28, 2021, https://www.nbcnews.com/health/health-news/chocolate-cravings-may-be-real-gut-feeling-flna1c9456552.
6. "1952 Nobel Prize": "The Nobel Prize for Physics 1952," the Nobel Prize (website), accessed November 29, 2021, https://www.nobelprize.org/prizes/physics/1952/summary/.
7. "behavior of hydrogen atoms": K. D. Stephan, "How Ewen and Purcell Discovered the 21-cm Interstellar Hydrogen Line," *IEEE Antennas and Propagation Magazine* 41, no. 1 (February 1999), accessed November 29, 2021, https://ieeexplore.ieee.org/document/755020.
8. "professions not your own": Martin Harwit, *Cosmic Discovery: The Search, Scope, and Heritage of Astronomy* (New York: Cambridge University Press, 2019).
9. "heartbeat via ultrasound": Healthline, "How Early Can You Hear Baby's Heartbeat on Ultrasound and by Ear?," accessed April 5, 2022, https://www.healthline.com/health/pregnancy/when-can-you-hear-babys-heartbeat.
10. "fifteen most religious states": Michael Lipka and Benjamin

Wormald, "How Religious Is Your State?," Pew Research Center, February 29, 2016, accessed December 12, 2021, https://www.pewresearch.org/fact-tank/2016/02/29/how-religious-is-your-state/?state=alabama.

11. "greatly restrict abortion": Guttmacher Institute, "Abortion Policy in the Absence of Roe," accessed December 12, 2021, https://www.guttmacher.org/state-policy/explore/abortion-policy-absence-roe.
12. "the death penalty": Death Penalty Information Center, "State by State," accessed December 12, 2021, https://deathpenaltyinfo.org/state-and-federal-info/state-by-state.
13. "three out of four Republican voters": Gallup, "Abortion Trends by Party Identification 1995-2001," accessed April 8, 2022, https://news.gallup.com/poll/246278/abortion-trends-party.aspx.
14. "a fetus thereafter until birth": Healthline, "Embryo vs Fetus: Fetal Development Week by Week," Healthline, accessed April 8, 2022, https://www.healthline.com/health/pregnancy/embryo-fetus-development.
15. "abort the babies": See, e.g., Laura Ingraham (@IngrahamAngle), Twitter, December 27, 2012, 9:47 a.m., accessed April 9, 2022, https://twitter.com/ingrahamangle/status/284309497294512128.
16. "5 million known pregnancies": Statista, "Number of Births in the United States from 1990 to 2019," accessed December 12, 2021, https://www.statista.com/statistics/195908/number-of-births-in-the-united-states-since-1990/. Note further: Reported births in the US for 2019 = 3.75 million. Add 630,000 medical abortions and at least 750,000 known spontaneous abortions, and you get 5.1 million pregnancies that year.
17. "medically aborted": Katherine Kortsmit et al., "Abortion Surveillance—United States, 2019," *Morbidity and Mortality Weekly Report* (*MMWR*) 70, no. 9 (November 26, 2021): 1–29, accessed December 12, 2021, https://www.cdc.gov/mmwr/volumes/70/ss/ss7009a1.htm.

18. "30 percent of all pregnancies": John P. Curtis, "What Are Abortion and Miscarriage?," eMedicine Health, accessed April 11, 2022, https://www.emedicinehealth.com/what_are_abortion_and_miscarriage/article_em.htm.
19. "physiological chauvinism": Rosemarie Garland Thomson, *Extraordinary Bodies: Figuring Physical Disability in American Culture and Literature* (New York: Columbia University Press, 1997).
20. "Captain on the bridge": Unpublished letter from Helen Keller to Captain von Beck, in the private collection of the author.
21. "legs, feet, and toes": Matt Stutzman, interviewed for *StarTalk Sports Edition*, August 2021, accessed November 24, 2021, https://www.youtube.com/watch?v=7NipfdwGTUs.
22. "Jahmani Swanson": Jahmani Swanson, interviewed for *StarTalk Sports Edition,* December 2021, accessed January 5, 2022, https://www.startalkradio.net/show/globetrotters-guide-to-the-galaxy/.
23. "influential people in the world": "The 2010 Time 100," *Time,* 2010, http://content.time.com/time/specials/packages/completelist/0,29569,1984685,00.
24. "until his death in 2018": Stephen Hawking, interviewed for *StarTalk*, March 14, 2018, accessed November 24, 2021, https://www.youtube.com/watch?v=TwaIQy0VQso.
25. "his own face in the mirror": Oliver Sacks, interviewed for *StarTalk*, "Are You Out of Your Mind?," accessed November 24, 2021, https://www.startalkradio.net/show/extended-classic-are-you-out-of-your-mind-with-oliver-sacks/.
26. "could not hear the music": Often attributed to nineteenth-century German philosopher Friedrich Nietzsche.
27. "but was never true": Christian Jarrett, *Great Myths of the Brain* (Hoboken, NJ: Wiley-Blackwell, 2014).
28. "no matter the stimulus": Daniel Graham, "You Can't Use 100% of Your Brain—and That's a Good Thing," *Psychology Today*, February 19, 2021, accessed November 21, 2021, https://

www.psychologytoday.com/us/blog/your-internet-brain/202102/you-cant-use-100-your-brain-and-s-good-thing.

29. "20 percent of our body's energy": Nikhil Swaminathan, "Why Does the Brain Need So Much Power?," *Scientific American*, April 29, 2008, accessed November 29, 2021, https://www.scientificamerican.com/article/why-does-the-brain-need-s/.
30. "unwelcome anomalies arise": American Museum of Natural History, "Brains," accessed November 27, 2021, https://www.amnh.org/exhibitions/extreme-mammals/extreme-bodies/brains; see also "Brain-to-Body Mass Ratio," Wikipedia, accessed November 27, 2021, https://en.wikipedia.org/wiki/Brain-to-body_mass_ratio.
31. "bottle in the street": "Genius Magpie," YouTube, accessed November 27, 2021, https://www.youtube.com/watch?v=xVSr22kqSOs.
32. "stars in the Milky Way Galaxy": Bradley Voytek, "Are There Really as Many Neurons in the Human Brain as Stars in the Milky Way?," *Brain Metrics* (blog), May 20, 2013, accessed November 29, 2021, https://www.nature.com/scitable/blog/brain-metrics/are_there_really_as_many/.
33. "Rubik's Cube in 0.25 seconds": "Why It's Almost Impossible to Solve a Rubik's Cube in Under 3 Seconds," *Wired*, accessed November 28, 2021, https://www.youtube.com/watch?v=SUopbexPk3A.
34. "traffic fatalities": Centers for Disease Control and Prevention, "Road Traffic Injuries and Deaths—a Global Problem," accessed March 3, 2022, https://www.cdc.gov/injury/features/global-road-safety/index.html.

CODA: LIFE AND DEATH

1. "born every second": "Number of Births," The World Counts, accessed December 19, 2021, https://www.theworldcounts.com/populations/world/births.
2. "two per second, on average": Worldometer, "World Population," accessed December 19, 2021, https://www.worldometers.info.

3. "people in the year 1900": Max Roser, Esteban Ortiz-Ospina, and Hannah Ritchie, "Life Expectancy," *Our World in Data*, 2013, last revised October 2019, accessed December 19, 2021, https://ourworldindata.org/life-expectancy.
4. "seven times longer than dogs": Dog life expectancy: 11–13 years. Human life expectancy: 75–90 years.
5. "Cretaceous-Tertiary (K-T) event": More recently referenced by paleontologists as the Cretaceous-Paleogene (K-Pg) event.
6. "size of Mount Everest": "Why Did the Dinosaurs Die Out?," History, March 24, 2010, updated June 7, 2019, accessed December 21, 2021, https://www.history.com/topics/pre-history/why-did-the-dinosaurs-die-out-1.
7. "life on Earth almost ended entirely": Hannah Hickey, "What Caused Earth's Biggest Mass Extinction?," *Stanford Earth Matters*, December 6, 2018, accessed December 19, 2021, https://earth.stanford.edu/news/what-caused-earths-biggest-mass-extinction#gs.ju3zsy.
8. "naturally occurs": "The Holocene Epoch," UC Museum of Paleontology, Berkeley, accessed December 17, 2021, https://ucmp.berkeley.edu/quaternary/holocene.php; see also Gerardo Ceballos, Paul R. Ehrlich, and Peter H. Raven, "Vertebrates on the Brink as Indicators of Biological Annihilation and the Sixth Mass Extinction," *Proceedings of the National Academy of Sciences* 117, no. 24 (June 1, 2020): 13596, https://www.pnas.org/doi/10.1073/pnas.1922686117; Daisy Hernandez, "The Earth's Sixth Mass Extinction Is Accelerating," *Popular Mechanics*, June 3, 2020, accessed December 19, 2021, https://www.popularmechanics.com/science/animals/a32743456/rapid-mass-extinction/.
9. "99.9 percent of them have gone extinct": "Roundtable: A Modern Mass Extinction?," *Evolution*, accessed December 17, 2021, https://www.pbs.org/wgbh/evolution/extinction/massext/statement_03.html.
10. "using many approaches": W. Kip Viscusi, "The Value of Life

in Legal Contexts: Survey and Critique" (originally published in *American Law and Economics Review* 2, no. 1 [Spring 2000]: 195–222), accessed December 18, 2021, https://law.vanderbilt.edu/files/archive/215_Value_of_Life_Legal_Contexts.pdf.

11. "Another sample calculation": Sarah Gonzalez, "How Government Agencies Determine the Dollar Value of Human Life," NPR, April 23, 2020, accessed December 18, 2021, https://www.npr.org/2020/04/23/843310123/how-government-agencies-determine-the-dollar-value-of-human-life.
12. "$233,000": Elyssa Kirkham, "A Breakdown of the Cost of Raising a Child," Plutus Foundation, February 2, 2021, accessed December 18, 2021, https://plutusfoundation.org/2021/a-breakdown-of-the-cost-of-raising-a-child/.
13. "twenty-one or younger": US Wings, "Vietnam War Facts, Stats and Myths," accessed June 2, 2022, https://www.uswings.com/about-us-wings/vietnam-war-facts/; see also National Archives, "Vietnam War U.S. Military Fatal Casualty Statistics," accessed June 2, 2022, https://www.archives.gov/research/military/vietnam-war/casualty-statistics.
14. "half million people per year": World Health Organization, "Malaria," December 6, 2021, accessed January 2, 2022, https://www.who.int/news-room/fact-sheets/detail/malaria.
15. "at least 10^{30} variations": Likely a gross underestimate. See *Quora* discussion, accessed July 6, 2022, https://www.quora.com/What-is-the-maximum-number-of-genetically-unique-individuals-that-human-genome-allows, accessed July 6, 2022.
16. "never even be conceived": A concept articulately conveyed in Richard Dawkins, *Unweaving the Rainbow: Science, Delusion and the Appetite for Wonder* (New York: Houghton Mifflin, 1998).
17. "I beseech you": Horace Mann, commencement address at Antioch College, Yellow Springs, Ohio, 1859.

INDEX

Entries in *italics* refer to illustrations.

ABOUT THE AUTHOR

Neil deGrasse Tyson is an astrophysicist and author of the #1 bestselling *Astrophysics for People in a Hurry*, among other books. He is the director of New York City's Hayden Planetarium at the American Museum of Natural History, where he has served since 1996. Dr. Tyson is also the host and cofounder of the Emmy-nominated popular podcast *StarTalk* and its spinoff *StarTalk Sports Edition*, which combine science, humor, and pop culture. He is a recipient of twenty-one honorary doctorates, the Public Welfare Medal from the National Academy of Sciences, and the Distinguished Public Service Medal from NASA. Asteroid 13123 Tyson, as well as the species of leaping frog *Indirana tysoni*, are each named in his honor. He lives in New York City.